GENETICS - RESEARCH AND ISSUES

# AN ESSENTIAL GUIDE TO CYTOGENETICS

# GENETICS - RESEARCH AND ISSUES

Additional books in this series can be found on Nova's website under the Series tab.

Additional e-books in this series can be found on Nova's website under the eBooks tab.

GENETICS - RESEARCH AND ISSUES

# AN ESSENTIAL GUIDE TO CYTOGENETICS

NAOMI NORRIS
AND
CARMEN MILLER
EDITORS

nova
science publishers
New York

**Library of Congress Cataloging-in-Publication Data**

ISBN: 978-1-53613-370-7

*Published by Nova Science Publishers, Inc. † New York*

# CONTENTS

# PREFACE

*An Essential Guide to Cytogenetics* explores the use of cytogenetic data for studies of frogs as well as the insights that hypotheses of phylogenetic relationships have added to this issue. The authors provide an overview of PcP190 satellite DNA, sex chromosome systems and B chromosomes found in Anura. This book also aim to establish the health effects of various activities and exposures by examining the levels of exposure and the biological effects resulting from the interaction between the organism and the chemical agent. Following this, a chapter is included which focuses on the complex karyotype issues in myelodysplastic diseases, leukemias, lymphomas and multiple myelomas as the authors see them in daily practice in their center. The authors investigate hybridization, suggesting that the viability of a hybrid between species with a chromosomal discrepancy may offer important hypotheses to explain the morphological, molecular, and cytogenetic diversity of the genus. Trichorhinophalangeal syndrome is analyzed, suggesting that in cases of cytogenetically-invisible alterations, parental FISH analysis as well as aCGH should be considered as part of the clinical baseline testing.

Chapter 1 - The order Anura currently encompasses over 6,800 amphibian species distributed in 56 families and is an interesting group for cytogenetic studies. Whereas some groups of species present conservative karyotypes, others are highly variable in diploid number or number/location of diverse chromosomal markers, such as nucleolus organizer regions, heterochromatic bands and specific satellite DNA sites. In some cases, karyotypic variation overcomes morphological diversification, turning cytogenetics into a useful tool for taxonomy. Special variation is observed with respect to sex chromosomes and sex determination systems, with both female and male heterogameties observed. Although most species already karyotyped do not show sex chromosome heteromorphism, distinct levels of differentiation are observed between the sex chromosomes of several species, which makes this group particularly interesting for studies of sex chromosome evolution. In this chapter, the authors explore the use of cytogenetic data for studies of frogs as well as the insights that hypotheses of phylogenetic relationships have added to this issue. In addition, the authors provide a brief review of PcP190 satellite DNA (with new data for the genus *Engystomops*), sex chromosome systems and B chromosomes found in Anura.

Chapter 2 - Humans are daily exposed to a variety of potentially harmful agents in the air they breathe, liquids they drink, food they eat and products they use. Long-standing evidence of the bond between health and environment has led to the recognition for the need of sustainable development. On the other hand, there is an increasing global awareness of the inevitable limits of individual health care and the need to complement such services with effective public health strategies. According to World Health Organization (WHO), cancer is a leading cause of death worldwide. Additionally, at least 200 000 people die every year from occupational or work-related cancers. Exposure assessment aims at prevention. Establishing the health effects of various activities and exposures requires information about the levels of exposure and the biological effects resulting from the interaction between the organism and the chemical agent. The resulting data provides a basis for designing effective prevention and mitigation strategies. Cytogenetic endpoints have long been applied in surveillance of human genotoxic exposure and early effects of genotoxic carcinogens. Assays measuring chromosomal aberrations (CAs), micronucleus (MN) and sister chromatid exchange (SCE) in lymphocytes are well-established techniques extensively used in human biomonitoring studies to assess DNA damage at the chromosomal level. The relevance of cytogenetic alterations as a cancer risk biomarker is further supported by epidemiologic data linking CAs and MN with cancer risk in human populations. Thus, the use of cytogenetic markers in human biomonitoring is of paramount importance due to its predictability regarding deleterious effects resulting from the exposure to environmental stressors.

Chapter 3 - The great era for classical cytogenetics started sixty years ago with the description of the twenty-three pairs of human chromosomes (Tjio and Levan, 1956) and the discovery of the Philadelphia chromosome, the first known chromosomal defect associated with a specific type of cancer (Nowell and Hungerford, 1960). Since then many recurrent chromosomal aberrations linked to specific hematological malignancies have been detected. The vast majority of these abnormalities can be detected by modern molecular genetic methods so it might seem there is no longer a need for microscopy. However, there is still a significant portion of hemato-oncologic patients for which the classical cytogenetic investigation is appropriate, i.e., cases with complex karyotypes. Complex karyotypes are characterized by the presence of three or more unrelated chromosomal aberrations co-existing in a single clone. Their occurrence is associated with adverse outcomes across the entire spectrum of hematologic malignancies. These aberrations are also a powerful diagnostic indicator for molecular targeted therapies, allogeneic stem cell transplantation or other generally more aggressive treatment strategies. Since the presence of complex karyotypes would be missed when using only molecular genetic methods, this highlights the irreplaceable role of classical cytogenetics as a first-tier analysis for the evaluation of complex structural chromosomal abnormalities in hemato-oncologic patients. Ideally, classical cytogenetics is then followed by more precise molecular genetic methods to identify specific chromosomal aberrations more deeply. Here the authors focus on the complex karyotype issues in the myelodysplastic diseases, leukemias, lymphomas and multiple myelomas as seen daily in their center.

Chapter 4 - Initially identified as *Astyanax schubarti* on the basis of morphological characteristics, its chromosomal analysis revealed a unique diploid number in the genus. With 42 chromosomes and the impossibility of homologous pairing, the karyotype of the individual was compared to a set of haploid complements of *Astyanax schubarti* (2n = 36 chromosomes) and *Astyanax fasciatus* (2n = 48 chromosomes). Natural hybrids are rare. The viability of a hybrid between species with such chromosomal discrepancy may offer important hypotheses to explain the morphological, molecular, and cytogenetic diversity of the genus.

Chapter 5 - There are three distinct subtypes of Trichorhinophalangeal syndrome (TRPS); TRPS type I, TRPS type II and TRPS type III. Features common to all three subtypes include sparse, slowly growing scalp hair, laterally sparse eyebrows, a bulbous tip of the nose (pear-shaped), and protruding ears. The diagnosis of TRPS is based in typical clinical and radiographic features, as well as in the identification of causative mutations in the *TRPS1* gene (TRPS type I and III) and in loss of functional copies of the *TRPS1* and EXT1 genes (TRPS type II). The mode of inheritance in TRPS I and III is autosomal dominant, however *de novo* deletions of *TRPS1* and EXT1 genes are the main defect of TRPS II. Parental balanced chromosomal rearrangements are an important cause of interstitial aberrations in TRPS. In cases of cytogenetically-invisible alterations, parental FISH analysis as well as aCGH should be considered as part of the clinical baseline testing. Treatment of clinical problems of TRPS types is mainly supportive and includes ectodermal and skeletal issues. The number of distinct syndromes as TRPS is rapidly increasing and their confirmation is necessary especially in cases that typical features are absent. Clinical geneticists should provide information for the families and advise them how to overcome problems.

In: An Essential Guide to Cytogenetics
Editors: N. Norris and C. Miller
ISBN: 978-1-53613-370-7


*Chapter 1*

# ANURAN CYTOGENETICS: AN OVERVIEW

***Cíntia Pelegrineti Targueta[1], Stenio Eder Vittorazzi[2], Kaleb Pretto Gatto[1], Daniel Pacheco Bruschi[3], Ana Cristina P. Veiga-Menoncello[1], Shirlei M. Recco-Pimentel[1] and Luciana Bolsoni Lourenço[1,*]***

[1]Department of Structural and Functional Biology, University of Campinas (Unicamp), Campinas, São Paulo, Brazil
[2]Department of Biological Sciences, Mato Grosso State University, Tangará da Serra, Mato Grosso, Brazil
[3]Department of Genetics, Federal University of Paraná (UFPR), Curitiba, Paraná, Brazil

## ABSTRACT

The order Anura currently encompasses over 6,800 amphibian species distributed in 56 families and is an interesting group for cytogenetic studies. Whereas some groups of species present conservative karyotypes, others are highly variable in diploid number or number/location of diverse chromosomal markers, such as nucleolus

[*] Corresponding Author Email: bolsoni@unicamp.br.

organizer regions, heterochromatic bands and specific satellite DNA sites. In some cases, karyotypic variation overcomes morphological diversification, turning cytogenetics into a useful tool for taxonomy. Special variation is observed with respect to sex chromosomes and sex determination systems, with both female and male heterogametes observed. Although most species already karyotyped do not show sex chromosome heteromorphism, distinct levels of differentiation are observed between the sex chromosomes of several species, which makes this group particularly interesting for studies of sex chromosome evolution. In this chapter, we explore the use of cytogenetic data for studies of frogs as well as the insights that hypotheses of phylogenetic relationships have added to this issue. In addition, we provide a brief review of PcP190 satellite DNA (with new data for the genus *Engystomops*), sex chromosome systems and B chromosomes found in Anura.

## 1. Cytogenetics, Taxonomy and Phylogenetic Inferences

In the last 15 years, knowledge of frog taxonomy and systematics has increased dramatically, boosted by the inclusion of DNA sequences in analyses. Phylogenetic relationships at the family level (Frost et al., 2006; Grant et al., 2006; Pyron & Wiens, 2011; Padial et al., 2014; Duellman et al., 2016; Feng et al., 2017) and especially in lower taxonomic levels (examples in Faivovich et al., 2012; Blotto et al., 2013; De Sá et al., 2014; Vacher et al., 2017; Chen et al., 2017) have been inferred, motivating several taxonomic revisions. The number of amphibian species descriptions has risen rapidly (Kohler et al., 2005), with over 100 species of frog described between January and November 2017 (based on to the list provided by Frost, 2017), but the species-level diversity of amphibians remains underestimated (see analysis provided by Giam et al., 2012; for examples of recent disclosure of still unnamed species, see Funk et al., 2012; Caminer et al., 2017; Chen et al., 2017, among others). In parallel, several species have been synonymized (examples in Guarnizo et al., 2012; Orrico et al., 2013, 2017).

In this scenario of frequent taxonomic changes and uncertainties, particular attention should be devoted to the identification of cytogenetic data available for frogs. There are several examples in the literature of species with cytogenetic data described or discussed under different names, which brings additional complexity when assembling specific cytogenetic data. That is the case, for example, for the craugastorid species currently known as *Strabomantis biporcatus*, whose karyotype was first described as belonging to *Eleutherodactylus maussi* (Schmid et al., 1992, 2002a), or the dendrobatid *Ranitomeya vanzolinii*, whose karyotype was first described by Bogart (1991) and assigned to *Dendrobates quinquevittatus* (a synonym junior of *Ameerega quinquevittatus*) (see comments in Grant et al., 2006 and Rodrigues et al., 2011). Taking into account the recent improvement in DNA barcoding of frogs (Vences et al., 2005a,b; Fouquet et al., 2007; Lyra et al., 2016), an approach that has become useful to minimize taxonomic problems is to obtain DNA sequences from the same individuals analyzed cytogenetically (Targueta et al., 2010; Medeiros et al., 2013; Veiga-Menoncello et al., 2014; Teixeira et al., 2016).

The increasing number of phylogenetic studies of frogs has also improved inferences of karyotypic evolution. The evolutionary comparison of karyotypes involves the inference of interspecific chromosomal homeologies and changes, which relies on the identification of many chromosomal markers. Therefore, the greater the number of chromosomal markers available for a representative sample of the taxonomic group of interest and its outgroup, the better the chances of obtaining informative data for the inference of karyotypical evolution. However, the chromosomal markers usually obtained in cytogenetic studies of frogs are scarce, with most of them being restricted to diploid number, chromosomal morphology (i.e., chromosome size and centromeric position), and location of heterochromatin and nucleolus organizer regions (NORs). Although some important phylogenetic groups were first recognized based on their diploid numbers, as the 30-chromosome hylids (currently the *Dendropsophus* genus – see Faivovich et al., 2005) and the *Sclerophrys regularis* group (known as the 20-chromosome toads – see Cunningham & Cherry, 2004 and references therein), the available chromosome markers have usually been insufficient for the proposition of substantial chromosomal evolutionary hypotheses. Accordingly, proper karyotypic evolutionary inferences have been facilitated by the tracking of chromosomal characters in cladograms inferred from other source of data, mostly DNA sequence matrices (examples in Lourenço et al., 2008, 2015; Cardozo et al., 2016). This approach involves the optimization of chromosomal characters based on the parsimony criterion, and has allowed for the recognition of a number of putative chromosomal synapomorphies as well as homoplastic characters, either concerning diploid chromosome number, chromosome morphology, NOR location or heterochromatic bands, as exemplified below.

### 1.1. Inferences of Evolutionary Changes in Diploid Number

Diploid chromosome number is easily coded as a phenotypic character and can be readily included in data matrices used for phylogenetic inferences (see example in Grant et al., 2006). The amount of missing data regarding diploid number of frogs, however, is still extensive, and most available inferences of transformations related to this character have derived from its analysis in the light of phylogenetic hypotheses inferred previously or from other sources of data. Some examples of chromosome number transformations already inferred for anurans are listed here.

1. Progressive reduction in diploid number from 22 to 20, 18 and 16 in the leptodactylid genus *Pseudopaludicola* (Veiga-Menoncello et al., 2014; Cardozo et al., 2016), with a parallel origin of a 20-chromosome karyotype in *Pseudopaludicola boliviana* (Cardozo et al., 2016).
2. Increase in chromosome number from 2n = 24 to 2n = 30 as a synapomorphy of the hylid genus *Dendropsophus* (Suárez et al., 2013; Faivovich et al., 2005).
3. Three independent changes in diploid number from the plesiomorphic 2n = 26 to 2n = 30, 2n = 34, and 2n = 22 in the alsodid genus *Alsodes* (Blotto et al., 2013).

4. Reduction from 2n = 24 to 2n = 22 due to a putative fusion of chromosomes 3 and 12 as a synapomorphy of the *Aplastodiscus albofrenatus* group (Hylidae), and independent change of the plesiomorphic condition 2n = 24 in the *Aplastodiscus albosignatus* group (sensu Berneck et al., 2016) (Gruber et al., 2012a; Berneck et al., 2016). Berneck et al. (2016) note that changes in diploid number in the *A. albosignatus* group are still ambiguous as the karyotypes of most species of this group are still unknown but consider the hypothesis of a former reduction to 2n = 20 and a sequential reduction to 2n = 18 in this group (as previously proposed by Gruber et al., 2012a).
5. Reduction from 2n = 22 to 2n = 20 as a synapomorphy of the Duovox clade of the leptodactylid genus *Engystomops* (Targueta et al., 2012).
6. Reduction from 2n = 22 to 2n = 20 as a synapomorphy of the *Sclerophrys regularis* group (Bufonidae), with a reversion to 2n = 22 in *Sclerophrys pardalis*, was inferred in the phylogenetic study conducted by Cunningham & Cherry (2004). The authors, however, emphasize the need of further analysis to verify this hypothesis of reversion as they could not statistically discard the alternative arrangement of *S. pardalis* as the sister species of the group composed of the 20-chromosome toads.
7. Increase from 2n = 24 to 2n = 28 in the hylid species *Pseudis cardosoi* as a result of two centric fission events (Busin et al., 2001; Aguiar-Jr. et al., 2007).
8. Reduction from 2n = 24 to 2n = 22 in the hylids *Phyllodytes edelmoi* and *Phyllodytes luteolus* (Gruber et al., 2012b). Because the interspecific relationships in *Phyllodytes* remain unknown as well as the diploid number of other species of this genus, it is not possible to evaluate if the diploid numbers of *P. edelmoi* and *P. luteolus* have resulted from the same evolutionary event or if this condition is shared with other *Phyllodytes* species. However, the

reduction in diploid number could be inferred considering that 2n = 24 is a putative synapomorphy of the subfamily Hylinae, as suggested by Faivovich et al. (2005) (Gruber et al., 2012b). Also based on this assumption, Gruber et al., (2007) inferred the reduction from 2n = 24 to 2n = 22 in the hylid species *Boana albopunctata*.

Although some rearrangements and chromosomes have been considered as possibly involved in some of the aforementioned cases (see discussions in Gruber et al., 2007, 2012a,b; Cardozo et al., 2016), only the diploid reduction observed in *Pseudis cardosoi* could be associated with an explicit proposal of chromosome rearrangements (Busin et al., 2001). In this case, based on chromosome morphology, C-bands and NOR site, Busin et al., (2001) inferred a reliable hypothesis of homeology between the chromosomes of *P. cardosoi* and its sister species *P. minuta* (see Aguiar-Jr. et al., 2007), according to which the bi-armed chromosomes 1 and 4 of *P. minuta* would be homeologous, respectively, to the telocentric chromosomes 6 and 7 and chromosomes 8 and 9 of *P. cardosoi*. Based on this hypothesis of homeology, and considering 2n = 24 as the plesiomorphic state for the group in analysis, Busin et al., (2001) suggested that chromosomes 6, 7, 8 and 9 of *P. cardosoi* have arisen from centric fissions of the same ancestral chromosomes that have led to chromosomes 1 and 4 of *P. minuta.*

Phylogenetic inferences have also assisted the analysis of changes in chromosome number that are derived from polyploidization. In the study of the diploid *Dryophytes chrysoscelis* and the tetraploid *Dryophytes versicolor* (as *Hyla versicolor*), the phylogenetic analyses supported the hypothesis that *D. versicolor* is possibly polyphyletic, as tetraploids were inferred to have arisen multiple times in this group (Ptacek et al., 1994; Holloway et al., 2006). Another interesting example is found in the genus *Xenopus*, in which independent and recurrent polyploidization events have been inferred (Evans et al., 2004, Evans, 2007, 2008).

## 1.2. Inferences Related to Chromosomal Morphology or Chromosomal Markers

In contrast to coding of diploid chromosome number, coding of other chromosome characters (such as chromosome morphology, chromosome markers or even chromosome rearrangements) is more complex and depends on reliable hypotheses of interspecific chromosome homeology. Usually, the description of frog karyotypes does not provide detailed characterizations of each chromosome, as the available information is typically limited to size, centromeric position (which allows chromosome classification as metacentric, submetacentric, subtelocentric, and telocentric, following the criteria proposed by Green & Sessions, 1991), and presence of C-bands and NORs. The paucity of informative markers prevents proper comparisons among karyotypes and as a result many karyotypes were described without any discussion of primary hypotheses of interspecific chromosome homeology, with chromosomes arranged by size and numbered in decreasing order. In addition, as different chromosomes of the same karyotype may present similar lengths and centromeric positions, this criterion of chromosome classification and karyogram organization has proved to be ineffective as it may cause errors in chromosome numbering, which adds further problems for the comparison of described karyotypes.

Measurement of chromosomes using specific software (as MicroMeasure – Reeves & Tear, 2000; IdeoKar – Mirzaghaderi & Marzangi, 2015; KaryoType – Altinordu et al., 2016; Drawid – Kirov et al., 2017) may help minimize huge discrepancies in karyotype descriptions, but it does not eliminate the problem. Taking measurement of centromeric regions or curved chromosome arms may be complicated and may vary from one research study to another. In addition, chromosome compaction degree varies amongst different metaphase cells, even in the same individual, and sometimes even between the homologues in a single cell, which may interfere in the calculation of chromosome relative length (RL) and arm ratio (AR) or centromeric index (CI). Special attention should be given to borderline values when ordering the chromosomes by size or when classifying the chromosomes with respect to their centromeric positions. The selection of good-quality metaphases, avoiding supercompacted, curved or crossed chromosomes, is of fundamental importance when measuring chromosomes and could minimize some measurement problems. Nevertheless, chromosome measurements are not accurate enough to allow for the usage of RL, AR and CI as powerful parameters for karyotypic comparisons. Although important, these parameters should not be the only or the first criterion for assessing chromosome pairing or organizing karyograms. It is not rare that the best hypothesis of arrangement of Giemsa-stained chromosome pairs in a karyogram proves to be wrong after the same chromosomes are C-banded or silver stained.

The sequential analysis of the same chromosomes submitted to different techniques is therefore indicated as a powerful approach for a better characterization of each chromosome of a karyotype. Sometimes different chromosomes that are very similar in morphology bear distinct markers and, in these cases, if the techniques that enable the detection of these markers are not applied sequentially, a proper interpretation of the karyotype is precluded. This is the case, for instance, for chromosomes 3 and 4 of the leptodactylid *Physalaemus ephippifer*. These chromosomes are very similar in size and centromeric position and may be easily misidentified in Giemsa-stained metaphases, but they may be differentiated by the presence of a pericentromeric DAPI-positive C-band in the short arm of chromosome 3, which is also detected by 5S rDNA and PcP190 satellite DNA probes (Nascimento et al., 2010; Vittorazzi et al., 2014a).

Therefore, when assessing interspecific chromosome homeologies for further identification of shared characters, chromosome morphology may help, especially when all (or most) chromosomes are similar between the karyotypes in comparison, but the analysis of specific chromosome markers is indispensable. Using this approach, some reliable primary homeologies may be recognized even between chromosomes identified by different numbers according to their sizes and, especially in these cases, the homeology hypotheses should be explained (for examples, see discussion about the NOR-bearing chromosomes found in *Paratelmatobius* and *Scythrophrys*, and in species of the hylid subfamily Scinaxinae – Lourenço et al., 2008 and Cardozo et al., 2011, respectively). In addition, when describing new karyotypes, it may be easier if the arrangement of chromosome pairs in karyograms reflects the homology hypotheses considered by the authors, even if the chromosome sizes suggest a distinct numbering. This type of decision was made when karyotypes of some species of *Physalaemus* were described (Tomatis et al., 2009; Vittorazzi et al., 2014b).

Once reliable chromosomal characters are recognized, interesting chromosomal synapomorphies, homoplasies, and evolutionary changes may be inferred. Some illustrative cases are found in the hylid Scinaxinae, which is currently composed of the genera *Sphaenorhynchus*, *Ololygon*, *Julianus*, and *Scinax* (Duellman et al., 2016), and in the leptodactylid genus *Physalaemus*. Cardozo et al., (2011), taking into account the phylogenetic relationships proposed by Faivovich (2002) and Faivovich et al., (2005), inferred the submetacentric morphology of chromosome pairs 1 and 2 and the NOR in chromosome pair 6 as three synapomorphies of *Ololygon* (referred as *Scinax catharinae* clade in Cardozo et al., 2011). In addition, Cardozo et al., (2011) also recognized the NOR in chromosome pair 11 of *Ololygon canastrensis* as a reversion to the plesiomorphic state of this character.

With respect to *Physalaemus*, by tracing cytogenetic characters in a cladogram inferred from DNA sequences, a telocentric chromosome pair 11 was inferred as a synapomorphy of the *Physalaemus signifer* clade, and this character was confirmed as homoplastic in relation to the telocentric chromosome 11 of *P. fernandezae* (Lourenço et al., 2015 and references therein), as formerly hypothesized by Tomatis et al., (2009). The phylogenetic relationships inferred for this genus also allowed for the recognition of a large C-band of the short arm of chromosome 3 as a synapomorphy of the clade composed of *Physalaemus biligonigerus*, *P. marmoratus* and *P. santafecinus*, and as a homoplasy in relation to the large C-band found in the short arm of chromosome 3 of *P. nattereri* (Lourenço et al., 2015 and references therein), as previously suspected (Vittorazzi et al., 2014b). Additionally, the interstitial C-band in the metacentric chromosome 5 was confirmed as a synapomorphy of the *P. cuvieri* species group (Lourenço et al., 2015 and references therein).

## 2. Cytogenetics as a Helpful Taxonomic Tool

Some groups of species present a high degree of karyotypic similarity. The bufonid species *Rhinella crucifer*, *R. icterica*, *R. jimi*, *R. rubescens*, and *R. schneideri* provide an evident example of this, as their karyotypes could not be distinguished by chromosome morphology or position of NOR and C-bands (Kasahara et al., 1996, as species of *Bufo*; Amaro-Ghilardi et al., 2008, as species of *Chaunus*). On the other hand, in some cases notorious interspecific differences are noted, and sometimes karyotypic variation overcomes morphological diversification, turning cytogenetics into a useful tool for taxonomy.

The finding of a new diploid number (2n = 30) in the genus *Alsodes*, for example, helped in the description of *Alsodes norae* (Cuevas, 2008). In the case of the hylid species *Dendropsophus nanus* and *D. sanborni*, which are very similar morphologically, cytogenetic analyses assisted in taxonomic identification, as their karyotypes could be easily distinguished by the number of telocentric chromosomes (the fundamental number is 52 in *D. nanus* and 50 in *D. sanborni* – Medeiros et al., 2003).

In some cases, conspicuous karyotypic divergences are observed between specimens assigned to a single species, supporting suspicions of the presence of distinct unnamed taxa. For instance, in the leptodactylid genus *Engystomops*, cytogenetic signatures were found for *E. freibergi* and *E. petersi*, and two additional karyotypes were described for specimens that belongs to distinct clades as inferred from DNA sequence analyses, which likely correspond to still unnamed species (Targueta et al., 2010; Funk et al., 2012).

The *Pseudopaludicola* genus gives other striking examples of interspecific karyotypic divergences, with the first reports of diploid number variation dating from the 1960s and 70s (see Veiga-Menoncello et al., 2014 for references). Despite this, for long time the cytogenetics of

*Pseudopaludicola* was underexplored. After obtaining well-characterized samples, some of them coming from topotype specimens, the diploid chromosome number could be confirmed for *P. falcipes* (2n = 22), *P. mineira* (2n = 22), *P. saltica* (2n = 22), *P. ameghini* (2n = 20), *P. ternetzi* (2n = 20), *P. canga* (2n = 18), and *P. mystacalis* (2n = 16), and some karyotypes had to be assigned to unnamed species (Duarte et al., 2010; Fávero et al., 2011). Three of these unnamed species were described latter as *P. facureae* (Andrade & Carvalho, 2012), *P. atragula* (Pansonato et al., 2014), and *P. motorzinho* (Pansonato et al., 2016). In addition, based on difference of chromosome number, *P. ameghini* was removed from synonymy with *P. mystacalis* (Fávero et al., 2011), which contributed for solving recurrent taxonomic conflicts. Based on how informative is the diploid number for *Pseudopaludicola*, the inclusion of karyotypic information in the description of new species may be helpful and has been adopted by some authors (see descriptions of *P. murundu* – Toledo et al., 2010; *P. jaredi* – Andrade et al., 2016; and *P. ibisoroca* – Pansonato et al., 2016).

# 3. How to Study Frog Chromosomes

Giemsa staining, C-banding, silver staining and *in situ* hybridization to reveal NORs, and, to a lesser extent, staining with base-specific fluorochromes (especially 4',6-diamidino-2-phenylindole – DAPI, chromomycin A3 – CMA3, and mythramycin – MM) are techniques commonly applied to the study of frog karyotypes. Unfortunately, informative markers are not easily obtained throughout euchromatic regions of anuran chromosomes, although very interesting data have been generated by replication banding using BrdU incorporation in some species (example in Schmid & Steinlein, 2015).

Given the importance of chromosomal markers for reliable interspecific karyotypic comparisons (as discussed in item 1.2), the prospect of new chromosome markers may be very helpful. In this context, besides the chromosome mapping of satellite DNA (as the PcP190 satellite

DNA – see item 4), the production of chromosome probes from microdissected or flow-sorted material arises as an interesting strategy. Although chromosome painting has been frequently used for cytogenetic studies of several animals and plants (Hassanane et al., 1998; Shibata et al., 1999; Karamysheva et al., 2002; Marchal et al., 2004; Rubtsov et al., 2004; Diniz et al., 2008; Teruel et al., 2009a; Teruel et al., 2009b; Wang et al., 2009; Henning et al., 2011; Vicari et al., 2011; Kawagoshi et al., 2012; Silva et al., 2016; Pita et al., 2017; Romanenko et al., 2017), few studies have employed this technique for anuran cytogenetics (Krylov et al., 2010; Gruber et al., 2014; Uno et al., 2015; Targueta et al., in press).

Comparative gene mapping is another valuable strategy as it may identify syntenic blocks that are conserved in different species. Uno et al., (2013), for example, based on the chromosome mapping of 60 genes, recognized homeologous chromosome groups between *Xenopus tropicalis* and the allotetraploid species *Xenopus laevis* and inferred the occurrence of inter- and intrachromosomal rearrangements during the evolution of these frogs that could not be inferred by other approaches.

Finally, the comparative analysis of genomes has emerged as a powerful tool, as it allows for the identification of specific regions (sex specific, for example), which can be further mapped onto chromosomes. This promising approach has greatly improved the studies of sex chromosomes, as discussed in item 5 of this chapter.

# 4. PcP190 Satellite DNA

## 4.1. Satellite DNA as Chromosome Markers

Satellite DNA (satDNA) comprises tandemly repetitive sequences in genomes and is mostly located in centromeric and telomeric regions (Charlesworth et al., 1994; Elder & Tuner, 1995; Richard et al., 2008; López-Flores & Garrido-Ramos, 2012). These sequences can be present in large quantity in genomes, and several distinct families of satDNA can coexist in the same genome (Slamovits & Rossi, 2002; Plohl et al., 2008,

2012). An important feature of satDNA repeats is their non-independent (concerted) evolution, which may lead to a high level of sequence homogeneity among the repeats of a satDNA family (Dover, 1982, 1986; review by Plohl et al., 2008). Repeat homogeneity is achieved by mechanisms of non-reciprocal sequence transfer (such as unequal crossover, gene conversion, rolling circle replication and reinsertion, and transposition), and higher sequence similarity is expected among adjacent repeats than among repeats from different arrays of the same chromosome, repeats positioned on homologous chromosomes and repeats on non-homologous chromosomes, consecutively (Plohl et al., 2008, 2012). Additionally, the size of the repetitive blocks may vary greatly in the same genome and between organisms, even those related to functional centromeres (Warburton et al., 2000; Henikoff et al., 2001; Plohl et al., 2012).

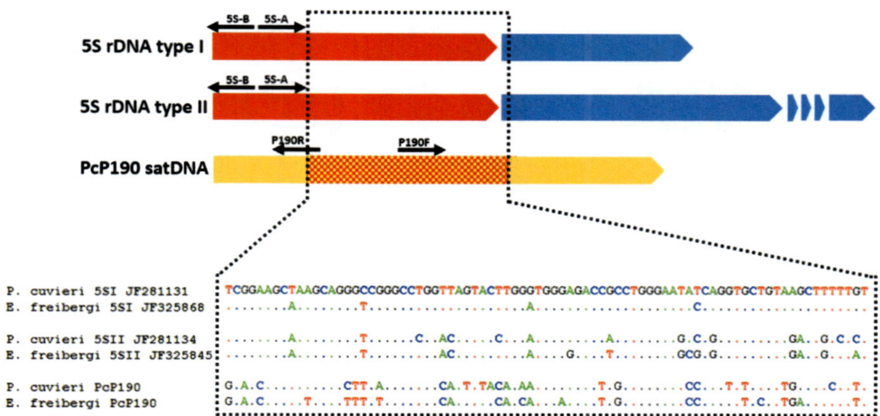


Figure 1. Comparison between type I and II 5S rDNA and PcP190 satDNA. Type I and type II 5S rDNA coding regions are shown in red, and non-transcribed regions (NTS) are shown in blue. Type I and type II 5S rDNA differ mostly in length and composition of NTS. An alignment of type I and type II 5S rDNA and PcP190 sequences from *P. cuvieri* and *E. freibergi* is also shown. GenBank accession numbers of the sequences are shown. Annealing regions of primers frequently used to isolate 5S rDNA (5S-A and 5S-B primers – Pendas et al., 1994) and PcP190 sequences are indicated (P190F and P190R – Vittorazzi et al., 2011).

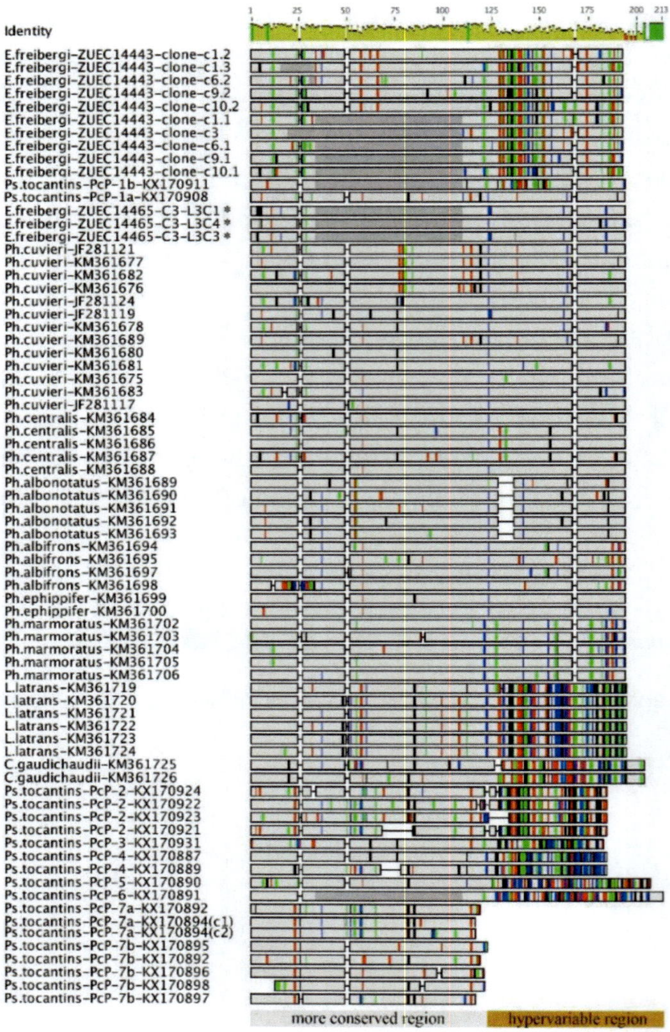


Figure 2. Alignment of all PcP190 sequences already described for anuran species. Only complete monomers are included, except for *Engystomops* and PcP-1b and PcP-6 sequences of *Pseudis tocantins*. The sequences are also identified by their GenBank accession numbers. Light gray boxes represent conservative sites and dark gray boxes are non-sequenced sites. Gaps are shown by white spaces and nucleotide polymorphisms, by colored boxes. Changes to thymine, cytosine, adenine, and guanine are represented by red, blue, green and black, respectively. Asterisks indicate the sequences obtained from microdissected chromosome 3 of *E. freibergi*. Note that these sequences are very similar to those obtained from *Physalaemus* species and to PcP-1a sequence of *Pseudis tocantins*.

Among evolutionarily closely related species, it is possible that the same satellite DNA family has different numbers of repeats, distributions and compositions, since the evolutionary dynamics of these repetitive sequences favors amplification and continuous deletion in the genome of organisms, so new families arise continuously by restructuring older sequences (Meštrović et al., 1998; Ugarkovic & Plohl 2002; Plohl et al., 2008).

Satellite DNA has been used as cytogenetic markers in comparative analysis of several organisms, such as insects, fish and mammals (Yamada et al., 2004; Saito et al., 2007; Acosta et al., 2007; Yoshimura et al., 2006; Cazaux et al., 2013; Matsubara et al., 2015; Ruiz-Ruano et al., 2016). The isolation and chromosome mapping of satellite DNA has also been employed in studies of some anurans (Table 1).

The satDNA PcP190 has proven to be widespread in frogs. It was first isolated from Leptodactylidae species (Vittorazzi et al., 2011) and has already been found in Hylodidae (Vittorazzi et al., 2014a) and Hylidae (Gatto et al., 2016) species to date. PcP190 satDNA is considered to be derived from 5S rDNA genes because of their high sequence similarity. Genetic similarities between the conserved region of Type I and II 5S rDNA and the corresponding regions in PcP190 were 70% and 66%, respectively, in *Physalaemus cuvieri* (Vittorazzi et al., 2011).

By comparing the nucleotide sequence of PcP190 fragments isolated from several frog species, two distinct regions can be recognized in the PcP190 repeats, one of which is more conserved, corresponding to the transcribing region of the 5S rDNA and one hypervariable region (Vittorazzi et al., 2014a, 2016; Gatto et al., 2016) (Figure 1). The hypervariable region varies both in size and nucleotide sequence and allows for the recognition of several different groups of sequences, including one group that is characterized by the absence of the hypervariable region (Gatto et al., 2016) (Figure 2).

## Table 1. Satellite DNAs described and chromosome mapped in species of Anura

| Family/Species | Satellite DNA (size of the repetitive unit) | Chromosome location | Reference |
|---|---|---|---|
| Alytidae | | | |
| *Discoglossus pictus* | pDS (288 bp) | 1 to 14 int | Odierna et al., 1999 |
| | Dp-sat1 (510 bp) | 8 to 14 cen | Amor et al., 2009 |
| | Dp-sat2 (143-148 bp) | 1 to 5 cen; 7 cen | Picariello et al., 2012 |
| | Dp-sat3 (170-177 bp) | 1 to 5 cen; 7 cen | Picariello et al., 2012 |
| Bufonidae | | | |
| *Bufotes viridis* | pBv (420 bp) | 3p per | Odierna et al., 2004 |
| Hylidae | | | |
| *Pseudis tocantins* | PcP190 (PcP-1a) (190 bp) | - | Gatto et al., 2016 |
| | PcP90 (PcP-1b) (?*) | Wq int | Gatto et al., 2016 |
| | PcP190 (PcP-2) (166-189 bp) | Wq int | Gatto et al., 2016 |
| | PcP190 (PcP-3) (181 bp) | Wq int | Gatto et al., 2016 |
| | PcP190 (PcP-4) (173-181 bp) | Wq int | Gatto et al., 2016 |
| | PcP190 (PcP-5) (204 bp) | Wq int | Gatto et al., 2016 |
| | PcP190 (PcP-6) (?*) | Wq int | Gatto et al., 2016 |
| | PcP190 (PcP-7) (107-121 bp) | Wq int | Gatto et al., 2016 |
| Hylodidae | | | |
| *Crossodactylus gaudichaudii* | PcP190 (201 bp) | - | Vittorazzi et al., 2014a |
| Leiopelmatidae | | | |
| *Leiopelma hochstetteri* | Lh1 (392 bp) | 10q int (Great Barrier Island, Big Omaha)/11q int, B1, B2 per (Mt Moehau; Tapu) | Zeyl & Green, 1992 |
| Leptodactylidae | | | |
| *Engystomops coloradorum* | PcP190 (unknown) | 5q per | This work |
| *Engystomops freibergi* | PcP190 (189-190 bp) | 3p per | This work |
| *Engystomops guayaco* | PcP190 (unknown) | 3p per; 3q per; 5q per | This work |
| *Engystomops "magnus"* | PcP190 (unknown) | 5p per; 5q per | This work |
| *Engystomops montubio* | PcP190 (unknown) | 3p per; 3q per | This work |
| *Engystomops petersi* | PcP190 (unknown) | 3p per; 3q per | This work |

| Family/Species | Satellite DNA (size of the repetitive unit) | Chromosome location | Reference |
|---|---|---|---|
| Leptodactylidae (continued) | | | |
| *Engystomops pustulatus* | PcP190 (unknown) | 1p per; 1q per; 3p per; 4q per | This work |
| *Engystomops randi* | PcP190 (unknown) | 3p per; 3q per; 5q per | This work |
| *Physalaemus albifrons* | PcP190 (190 bp) | 3p per | Vittorazzi et al., 2014a |
| *Physalaemus albonotatus* | PcP190 (183 bp) | 3p per | Vittorazzi et al., 2014a |
| *Physalaemus centralis* | PcP190 (190 bp) | 1 to 5 cen; 8 cen; 10 cen; 3p; 4p per | Vittorazzi et al., 2014a |
| *Physalaemus cicada* | PcP190 (200 bp) | 1 cen | Vittorazzi et al., 2016 |
| *Physalaemus cuvieri* - BA | PcP190 (190 bp) | 1 to 5cen; 3p per | Vittorazzi et al., 2011 |
| *Physalaemus cuvieri* - MG | PcP190 (190 bp) | 1 to 7 cen; 9 to 11 cen | Vittorazzi et al., 2014a |
| *Physalaemus cuvieri* - MS | PcP190 (190 bp) | 1 to 7 cen; 9p to 11p per | Vittorazzi et al., 2014a |
| *Physalaemus cuvieri* - RS | PcP190 (190 bp) | 1 to 5 cen | Vittorazzi et al., 2014a |
| *Physalaemus cuvieri* – PB | PcP190 (190 bp) | 1 to 11cen | Vittorazzi et al., 2014a |
| *Physalaemus cuvieri* – TO | PcP190 (190 bp) | 1 to 3 cen; 4p to 7p per; 10p per | Vittorazzi et al., 2014a |
| *Physalaemus ephippifer* | PcP190 (190 bp) | 3p per; Wq int | Vittorazzi et al., 2014a |
| *Physalaemus kroyeri* | PcP190 (190 bp) | 1 cen; 3p per | Vittorazzi et al., 2016 |
| *Physalaemus marmoratus* | PcP190 (190 bp) | - | Vittorazzi et al., 2014a |
| *Leptodactylus latrans* | PcP190 (192 bp) | - | Vittorazzi et al., 2014a |
| Pipidae | | | |
| *Xenopus laevis* | Xl32 (79 bp) | 1 to 9Lpq ter, 1 to 9Spq ter | Spohr et al., 1981; Schmid & Steinlein, 2015 |
| | Satellite 1 (741 bp) | - | Lam & Carroll, 1983a |
| | PTR-2 (388 bp) | - | Lam & Carroll, 1983b |
| | OAX (752 bp) | - | Ackerman, 1983 |
| | REM 2 (487 bp) | Disperse in almost all chromosomes | Hummel et al., 1984 |
| | REM 3 (463 bp) | 1Lp per | Hummel et al., 1984 |
| Ranidae | | | |
| *Glandirana rugosa* | 41-REL (41 bp) | 1pq to 13pq ter | Suda et al., 2011 |
| | 31-REL (31 bp) | 1pq to 13pq ter; Wq per | Suda et al., 2011 |
| *Lithobates catesbeianus* | RcS1 (360 bp) | - | Wu et al., 1986 |
| | RcS2 (550 bp) | - | Wu et al. 1986 |

**Table 1. (Continued)**

| Family/Species | Satellite DNA (size of the repetitive unit) | Chromosome location | Reference |
|---|---|---|---|
| Ranidae (continued) | | | |
| *Pelophylax esculentus* | RrS1 (292 bp) | 1 to 5 cen | Ragghianti et al., 1995 |
| | Rana/PolII (227-235 bp) | 1 to 13 disp | Ragghianti et al., 1999 |
| *Pelophylax lessonae* | RrS1 (292 bp) | 1 to 13 cen | Ragghianti et al., 1999 |
| | Rana/PolII (227-235 bp) | 1 to 13 disp | Ragghianti et al., 1999 |
| *Pelophylax ridibundus* | RrS1 (292 bp) | 1 to 13 cen | Ragghianti et al., 1995 |
| | Rana/PolII (227-235 bp) | 1 to 13 disp | Ragghianti et al., 1999 |
| *Rana dalmatina* | S1a (494 bp) | 1 per; 2 per, 3p to 5p per, 9q to 13q per | Feliciello et al., 2006 |
| | S1b (332 bp) | 1 per; 2 per, 3p to 5p per, 9q to 13q per | Feliciello et al., 2006 |
| *Rana graeca* | S1b (280 bp) | 1 per, 4p and 5p per, 8q per | Picariello et al., 2002 |
| *Rana holtzi* | S1a (476 bp) | - | Picariello et al., 2016 |
| *Rana italica* | S1a (494 bp) | 1 to 13 cen | Cardone et al., 1997 |
| | S1b (280 bp) | 1 to 13 cen | Cardone et al., 1997 |
| *Rana macrocnemis* | S1a (476 bp) | - | Picariello et al., 2016 |
| *Rana tavasensis* | S1a (476 bp) | - | Picariello et al., 2016 |
| *Rana temporaria* | S1a (476 bp) | - | Feliciello et al., 2005 |

q = long arm; p = short arm; cen = centromere; int = interstitial position; per = pericentromeric position; ter = terminal position in the chromosome; disp: disperse in the chromosome; * Only incomplete monomers were sequenced.

While only one type of PcP190 sequence (Figure 2) was observed among the PcP190 fragments isolated from several species of *Physalaemus*, a great diversity of this satDNA was found in the hylid *Pseudis tocantins* (PcP-1 to PcP-7 sequence groups) (Figure 2; Table 1 and references therein).

Different types of sequences were found adjacent to one another in the *P. tocantins* genome, supporting the occurrence of rearrangements between the distinct groups, although the homogenization process was less effective in this genome than in the *Physalaemus* genomes (Gatto et al., 2016). Based on the similarity between the coding region of 5S rDNA and the more conserved region of PcP190, and considering the great diversity among the non-transcribed regions of 5S rDNA, Gatto et al., (2016) discuss the possibility that the different groups of PcP190 sequences may have arisen from illegitimate recombination events between PcP190 satDNA and variants of the 5S rDNA. It should also be considered that the lower sequence variation in a specific region of the PcP190 satDNA repeats could be explained by differential selective pressures (Vittorazzi et al., 2014a; Gatto et al., 2016). In this context, it is worth noting that the same type of PcP190 sequences found in the *Physalaemus* species was observed in the *Pseudis* species as well (Figure 2; Gatto et al., 2016).

Clusters of PcP190 satDNA have been located at the centromeric/pericentromeric regions of some autosome pairs in seven *Physalaemus* species (*P. albifrons*, *P. albonotatus*, *P. centralis*, *P. cicada*, *P. cuvieri*, *P. ephippifer*, and *P. kroyeri*), raising the hypothesis that this satDNA could be involved in centromere biology in the studied species (Vittorazzi et al., 2014a, 2016); however, this has not been tested to date. In *Pseudis tocantins*, PcP190 satDNA was mapped exclusively to the large heterochromatic band in the long arm of the W chromosome, suggesting that this satDNA may have played an important role in the evolutionary differentiation of the sex chromosomes in this case. A differential distribution of PcP190 sequences was also observed between the sex chromosomes Z and W of *P. ephippifer*, although in this species chromosome 3 also bears a PcP190 site (Vittorazzi et al., 2014a).

## 4.2. PcP190 satDNA in *Engystomops*: Sequences and Chromosome Locations

The Neotropical genus *Engystomops* encompasses two clades, Edentulus and Duovox, which comprise the species located in the east and west side of the Andean Mountain range, respectively (Ron et al., 2006; 2010). The Edentulus clade is composed of *Engystomops freibergi*, *E. petersi*, *E. pustulosus* and two Confirmed Candidate Species not yet formally described, *E. "magnus"* and *E. "selva"* (Funk et al., 2012; Trillo et al., 2017). The Duovox clade includes *E. randi*, *E. montubio*, *E. guayago*, *E. coloradorum*, *E. pustulatus*, and *E. puyango*. A diploid number reduction from 2n = 22 to 2n = 20 was considered as a synapomorphy of the Duovox clade, although the chromosome rearrangements involved in this event are not known (Targueta et al., 2012).

PcP190 satDNA is present in the genus *Engystomops* and PcP190 sequences could be isolated from the genome and from microdissected chromosome 3 of *Engystomops freibergi* (Figure 2). When compared to the conserved regions of the PcP190 sequences found in the leptodactylids *Physalaemus*, *Leptodactylus latrans*, *Crossodactylus gaudichaudii*, and the hylid *Pseudis tocantins*, the conserved region of the complete PcP190 monomers of *E. freibergi* showed from 83% to 88% of similarity (Table 2). Three of the *E. freibergi* PcP190 sequences are highly similar to the PcP190 sequences of *Physalaemus* and PcP-1a sequences of *Pseudis*, even with respect to the hypervariable region (Figure 2). The remaining ten sequences, however, differ from those sequences in a 26 nucleotide-region at the beginning of the hypervariable region (Figure 2).

All species from the Edentulus clade showed a highly similar hybridization pattern with the PcP190 probe. PcP190 satDNA was mapped to a pericentromeric region of the the short arm of the chromosome identified as 3 in *E. freibergi* and *E. petersi* and identified as 5 in *Engystomops "magnus"* (Figure 3A-C). In addition, clusters of PcP190 were also detected near the centromere on the long arm of chromosomes 3 and 5 of *E. petersi* and *E. "magnus,"* respectively (Figure 3B,C).

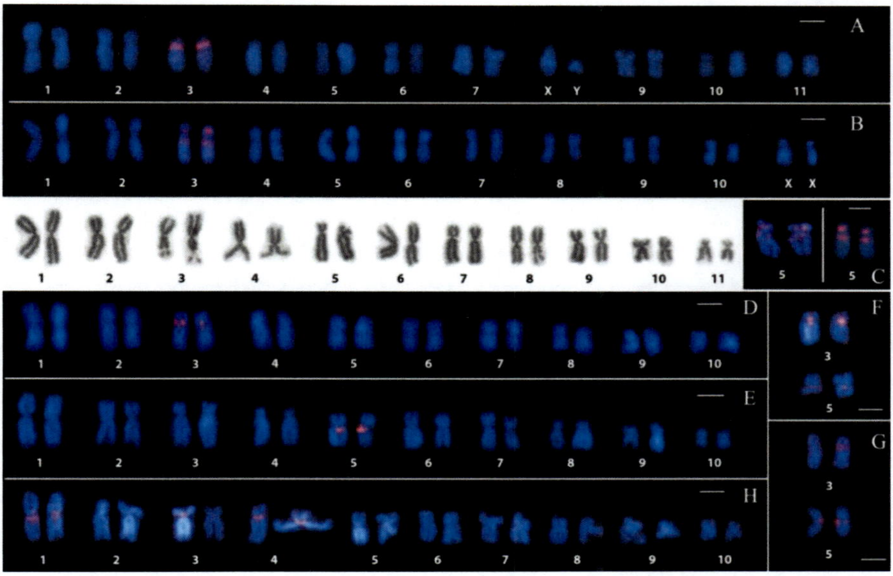

Figure 3. Chromosome mapping of PcP190 sequences in *Engystomops* species. In FISH essays, PcP190 probes were labeled with dUTP-digoxigenin and detected using anti-digoxigenin coupled to rhodamine, and the chromosomes were stained with DAPI. **A**. *E. freibergi*. **B**. *E. petersi*. **C**. *E. "magnus"*. **D**. *E. montubio*. **E**. *E. coloradorum*. **F**. *E. randi*. **G**. *E. guayaco*. **H**. *E. pustulatus*. In **C**, the karyotype of *E. "magnus"* is stained with Giemsa for a better visualization of chromosome morphology, and chromosome pair 5 from two metaphases hybridized with PcP190 probe are shown. Note the similarity between chromosome 5 of *E. "magnus"* (**C**) and chromosome 3 of *E. freibergi* (**A**) and *E. petersi* (**B**) with respect to morphology and location of PcP190 sites. See discussion about the possible homeology between pair 11 of *E. "magnus"* and sex chromosome pairs of *E. freibergi* and *E. petersi* in Targueta et al., (2010). Bar: 4 μm.

Chromosome pairs 3 of *Engystomops freibergi* and *E. petersi* are very similar in morphology but they deeply differ from pair 3 of "*E. magnus*". Based on classical cytogenetic methods, the homeology between pair 3 of *E. freibergi* and *E. petersi* and pair 5 of "*E. magnus*" was proposed

(Targueta et al., 2010), which was corroborated by the presence of a 5S rDNA site on the pericentromeric region of the short arm of these chromosomes (Rodrigues et al., 2012). Here we provide additional data to corroborate this homeology hypothesis, as chromosomes 3 of *E. freibergi* and *E. petersi* and chromosome 5 of *E. "magnus"* share the presence of PcP190 sites on both arms.

**Table 2. Genetic similarity (in percentage) between the conserved regions of the complete PcP190 sequences isolated from *Engystomops freibergi***

| | 1 | 2 | 3 | 4 | 5 | 6 | 7 | 8 |
|---|---|---|---|---|---|---|---|---|
| **1.** *E. freibergi* sequences | | | | | | | | |
| **2.** *Physalaemus* and *Pseudis* PcP-1 sequences | 87.63 | | | | | | | |
| **3.** *Pseudis* PcP-2 sequences | 78.63 | 83.44 | | | | | | |
| **4.** *Pseudis* PcP-3 sequences | 85.92 | 89.05 | 80.69 | | | | | |
| **5.** *Pseudis* PcP-4 sequences | 85.95 | 90.24 | 84.29 | 88.74 | | | | |
| **6.** *Pseudis* PcP-5 sequences | 84.21 | 89.77 | 82.18 | 86.44 | 89.28 | | | |
| **7.** *Pseudis* PcP-7 sequences | 82.77 | 87.50 | 88.45 | 84.19 | 87.65 | 86.94 | | |
| **8.** *C. gaudichaudii sequences* | 84.74 | 89.33 | 83.81 | 85.17 | 89.11 | 88.56 | 86.94 | |
| **9.** *L. latrans* sequences | 84.60 | 87.21 | 82.12 | 84.46 | 87.67 | 85.31 | 84.55 | 87.01 |

Source: (Present work) and those obtained from species of *Physalaemus* (Vittorazzi et al., 2011, 2014a, 2016), *Crossodactylus gaudichaudii* (Vittorazzi et al., 2014a), *Leptodactylus latrans* (Vittorazzi et al., 2014a), and *Pseudis tocantins* (Gatto et al., 2016), grouped according to their hypervariable region. Only conserved regions of complete monomers were considered in this analysis. PcP-1b and PcP-6 sequences of *Pseudis tocantins* were not included because no complete monomer is available

In contrast to the Edentulus clade, the karyotypes of the species of the Duovox clade are more similar to one another with respect to chromosome morphology, NOR position and heterochromatin distribution (Targueta et al., 2012) and they diverged more in relation to the location of PcP190 (Figure 3D-H). In the karyotypes of *E. montubio* (Figure 3D), *E. randi* (Figure 3F) and *E. guayaco* (Figure 3G), the PcP190 probe detected a pericentromeric region in both arms of chromosome 3. In the karyotypes of *E. randi* and *E. guayaco* (Figure 3F-G), the probe also mapped one pericentromeric region on the long arm of chromosome 5. In contrast, the accumulation of PcP190 sequences was detected only on a pericentromeric region on the long arm of chromosome 5 in *E. coloradorum* (Figure 3E), whereas in *E. pustulatus* the PcP190 probe detected the pericentromeric region on both arms of chromosome 1, a pericentromeric region on the short arm of chromosome 3 and a pericentromeric region on the long arm of chromosome 4 (Figure 3H).

Compiling all of the information regarding 5S rDNA and PcP190 sites on the karyotypes of *Engystomops* species (Figure 4), we can note that, at least in *E. freibergi*, *E. petersi*, and *E. "magnus"*, the type I 5S rDNA co-localizes with PcP190 clusters on the short arm of chromosome 3 (Rodrigues et al., 2012), as was also observed in *Physalaemus cuvieri* (Vittorazzi et al., 2011) and *P. ephippifer* (Nascimento et al., 2010; Vittorazzi et al., 2014a). This observation may contribute to the proposal that the origin of PcP190 satDNA occurred in an ancestral chromosome carrying 5S rDNA. Heteromorphic sex chromosomes X and Y are observed in *Engystomops freibergi* and *E. petersi*, and possibly heteromorphic Z and W chromosomes occur in *E. coloradorum* (Targueta et al., 2010; 2012), but none of them had clusters of PcP190 satellite DNA revealed by FISH (fluorescent *in situ* hybridization). In these cases, sex chromosome differentiation did not involve this satDNA, in contrast to what could be inferred for *Pseudis tocantins* (Gatto et al., 2016) and *P. ephippifer* (Vittorazzi et al., 2014a).

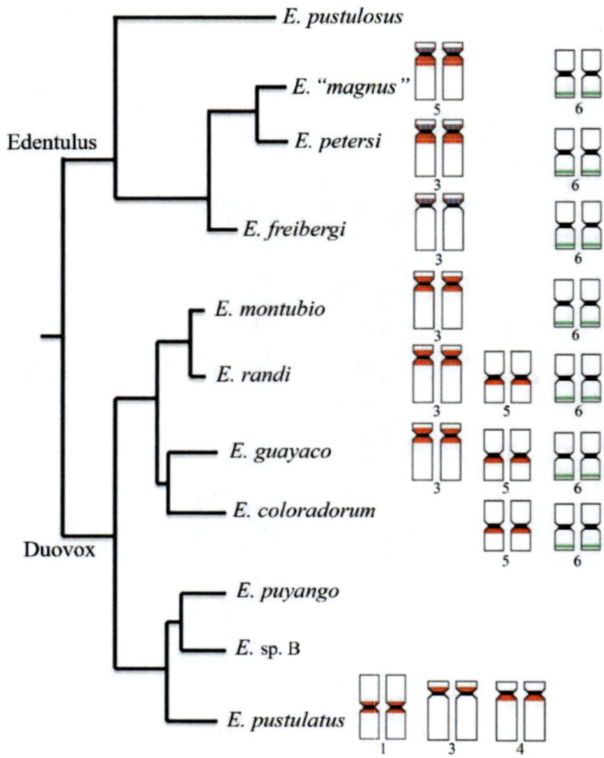


Figure 4. Chromosomal sites of PcP190 satDNA and 5S rDNA in *Engystomops*. The phylogenetic relationships in this genus are shown in the cladogram derived from the phylogenies inferred by Ron et al., (2010) and Funk et al., (2012) and ideograms show the hybridization sites of the PcP190 probe (red), type I 5S rDNA probe (blue; data obtained from Rodrigues et al., 2012) and type II 5S rDNA probe (green; data obtained from Rodrigues et al., 2012 and unpublished data) in each species. Note that type I 5S rDNA and PcP190 signals co-localize in *E. freibergi*, *E. petersi* and *E. "magnus"*. Because chromosomes 5 and 6 from species of the Duovox clade are very similar in size and centromeric position, simultaneous hybridization with the type II 5S rDNA probe and the PcP190 probe was needed to ensure that these probes detected distinct chromosomes. Based on a combined analysis of the phylogenetic and cytogenetic data, we may infer that the PcP190 cluster of chromosome 3 was lost in *E. coloradorum*. With respect to the PcP190 site of chromosome 5, one possible interpretation is that this PcP190 cluster was lost in *E. montubio* after having arisen in the common ancestor of *E. montubio*, *E. randi*, *E.guayaco*, and *E. coloradorum*. However, another hypothesis equally parsimonious is that a PcP190 cluster has arisen independently in chromosome 5 of *E. randi* and chromosome 5 of the common ancestor of *E. guayaco* and *E. coloradorum*.

# 5. Sex Chromosomes in Anura

## 5.1. Sex Determination in Anurans: More Questions than Answers

Genetic sex determination is a widely observed phenomenon in nature, mainly in animals and some plants, but only few genes are known for this function. In placental and marsupial mammals, the sex determination gene (SDG) is the gene *SRY* (sex determination region Y), a gene from the *SOX* gene family and a paralogue of the gene *SOX3* (Berta et al., 1990; Sinclair et al., 1990; Goodfellow & Lovell-Badge, 1993; Graves, 2008). In monotreme mammals, the candidate SDG is the gene *AMH*, which is located on the $Y_5$ chromosome (Cortez et al., 2014). For birds, characterized by female heterogamety, the SDG is the *DMRT1* (doublesex and mab-3-related transcription factor 1) gene, which is present on chromosome Z but absent on chromosome W, so that sex determination in avian species occurs by dosage of this gene (Nanda et al., 2000; Nanda et al., 2008; Smith et al., 2009). For non-avian reptiles that possess genetic sex determination, no candidate SDG is known to date. For fish species, there are two SDGs already characterized: the *DMY* gene, found in *Oryzias latipes*, a paralogue of the gene *DMRT1*; and the *AMHR2* gene, found in *Takifugu rubripes* (Paul-Prasanth et al., 2006; Matsuda et al., 2007; Graves & Peichel 2010).

For frogs, the only known SDG is the gene *DM-W*, a paralogue of *DMRT1* found on the W chromosome of *Xenopus laevis* (Yoshimoto et al., 2008; Mawaribuchi et al., 2017). *DM-W* probably arose from a duplication of *DMRT1* after the allotetraploidization event that occurred in the *X. laevis* lineage (Mawaribuchi et al., 2017). The gene *DM-W* is present in species closed related to *X. laevis*, such as *X. clivii*, *X. gilli*, *X. largeni* and *X. pygmaeus*, but it is absent in other species of the genus *Xenopus*, as in *X. tropicalis* and *X. muelleri* (Bewick et al., 2011). However, there is no evidence that *DM-W* is the SDG in all *Xenopus* species that possess it (Bewick et al., 2011). Recent studies identified sex-linked sequences in *X. borealis* that do not correspond to the *DM-W* gene and also inferred this species as the sister taxon of *X. clivii*. Because *DM-W* is present in *X. clivii*, and based on interspecific phylogenetic relationships, Furman & Evans (2016) suggested that *DM-W* arose as a sex-determining trigger in the ancestor of the group that includes *X. borealis*, *X. clivii*, *X. laevis*, *X. largeni*, and *X. allofraseri* and that *X. borealis* bears a new sex determination system that is derived with respect to the *DM-W*-based system.

In the ranid *Glandirana rugosa*, the *Sox3* gene was mapped to the sex chromosomes and has been considered as a candidate gene for sex determination, but its function in not yet elucidated (Uno et al., 2008; Miura & Ogata, 2013). Other candidate sex-determining gene is the *Dmrt1* found in some tree frog species (Dufresnes et al., 2015 and references therein).

These interesting findings concerning sex determination in frogs make evident how complex and diverse is this. Further studies on this issue will be certainly very elucidating as different sex determination systems, either with female and male heterogameties, are present in this group.

## 5.2. Sex Chromosome Differentiation

In organisms with genetic sex determination, the sex chromosomes harbor the sex determination gene that will trigger gonad development, in such a way that the heterogametic sex may be the male, such as in mammals (female: XX/male: XY), or female, such as in birds (male: ZZ/female: ZW) (for review, see Graves, 2008 and Schartl et al., 2016). Although sex chromosome differentiation has specific particularities in distinct lineages, some steps are common among all animals and plants that possess sex chromosomes, with the first one being the acquisition of a sex determination locus and the arrest of recombination in that region (Muller, 1964; Ohno, 1967; reviews in Graves, 2006, 2008). Suppression of recombination prevents breakage in sex determination linkage, allowing for the fixation of mutations on the chromosome exclusive of heterogametic sex (Y or W) (for review, see Graves, 2008 and Bergero &

Charlesworth, 2009). Accumulation of heterochromatin by the amplification/invasion of repetitive DNA elements is a common consequence of this process, and it may also enhance the arrest of recombination between X/Z and Y/W, leading to higher levels of heteromorphism of these chromosomes (Singh et al., 1976, 1980; Charlesworth et al., 2005; Steinemann & Steinemann, 2005). In the final stages of sex chromosome differentiation, Y/W chromosomes may undergo a degeneration process, reducing in size and harboring only few genes involved with sex differentiation and/or reproduction (see Steinemann & Steinemann, 2005; Graves, 2008).

In anurans, both female and male heterogameties are present and the occurrence of heteromorphic sex chromosomes is rare. To date, among the already karyotyped anuran species, 21 species with XX/XY and 24 species with ZZ/ZW systems present heteromorphic sex chromosomes (see review by Schmid et al., 2012; Saba & Tripathi, 2014; Patawang et al., 2014; Sangpakdee et al., 2017). In addition, heteromorphic sex chromosomes were already found in multiple sex chromosome systems (derived from Y-autosome fusions) in species of the genera *Pristimantis* and *Strabomantis* (family Craugastoridae) and in *Eleutherodactylus cavernicola* (family Eleutherodactylidae) (Schmid et al., 1992, 2002a, 2003, 2010). Another uncommon sex chromosome system occurs in *Leiopelma hochstetteri*, as females has a supernumerary chromosome, denoting a 00/0W sex determination system (Green, 1988).

In addition to the diversity in sex chromosome systems, a high diversity is also observed with respect to the level of sex chromosome heteromorphism. In some cases, sex chromosome heteromorphism can be revealed only after employing specific techniques. In some anuran species, for example, the sex chromosomes can only be distinguished by differential replication patterns (Schempp & Schmid, 1981; Odierna et al., 2007). The sex chromosomes of *Xenopus laevis* can be differentiated cytogenetically only by mapping the *DM-W* gene onto chromosome W by *in situ* hybridization (Yoshimoto et al., 2008). The sex chromosomes of the bufonid *Rhinella marina* were identified by CGH (Comparative Genomic Hybridization) (Abramyan et al., 2009), and some species of the genus

*Hyla* had a XX/XY sex determination system recognized only by linkage map analysis (Berset-Brändli et al., 2007). On the other hand, some species present highly differentiated sex chromosomes that differ both in morphology and molecular constitution. In *Pristimantis euphronides* and *Pristimantis shrevei*, for example, the Z chromosome is smaller than the W chromosome, which is almost entirely heterochromatic (Schmid et al., 2002a). In *Pseudis tocantins*, the submetacentric W chromosome is larger than the metacentric Z chromosome because of the presence of an amplified heterochromatin in Wq enriched for PcP190 satellite DNA (Busin et al., 2008; Gatto et al., 2016). In addition, these sex chromosomes differ with respect to the NOR, which is interstitial in Zq and pericentrimeric in Wq (Busin et al., 2008).

In most cases, however, intermediate stages of sex chromosome differentiation are noted. In some species, sex chromosomes are similar in size and centromeric position, but differ in heterochromatin accumulation and can be easily identified by C-banding (as in *Proceratophrys boiei* - Ananias et al., 2007, Amaro et al., 2012; in *Eupsophus roseus* - Iturra & Veloso, 1989; in *Gastrotheca pseutes* - Schmid et al., 1990; among several others). In other cases, sex chromosomes that are similar in size, centromeric position and heterochromatin amount can be distinguished by NOR location (as in *Dryophytes femoralis* – *Hyla femoralis* in Schmid & Steinlein, 2003; and possibly in *Engystomops coloradorum* (Targueta et al., 2012). There are also cases in which sex chromosomes show similar sizes but differ with respect to centromeric position (as in *Eleutherodactylus johnstonei* – Schmid et al., 2010; and in *Pseudopaludicola saltica* – Duarte et al., 2010) or with respect to centromeric position and additional features, such as C-band distribution (as in *Eupsophus miguelli* and *Eupsophus insularis*, whose X chromosome is telocentric and possesses centromeric heterochromatin, while the Y chromosome is metacentric and lacks the centromeric positive C band – Iturra & Veloso, 1989; Cuevas & Formas, 1996). Differences in size between X/Z or Y/W have also been reported for several other species (reviewed by Schmid et al., 2012).

Because of the presence of different levels of sex chromosome heteromorphism, the frogs are an interesting group for studies on sex chromosome differentiation.

## 5.3. Sex Chromosome Turnover and Occasional Recombination between Sex Chromosomes

Transition between female and male heterogamety is very common in amphibians and may be observed even within the same genus. Evans et al., (2012), using tools for ancestral character state reconstruction and a molecular phylogeny proposed by Pyron & Wiens (2011), showed that there is no support to infer whether the ancestral heterogamety in frogs (or even in salamanders and amphibians) was female or male but they expanded the hypothesis first proposed by Hillis & Green (1990) that amphibians experienced several XX/XY↔ZZ/ZW transitions. Over 30 cases of sex chromosome turnover in amphibians, including evolution of novel systems and reversals to ancestral conditions, were estimated by Evans and colleagues.

An interesting example of sex system transition is observed in *Glandirana rugosa*, as XX/XY and ZZ/ZW sex determination systems occur in geographically separated populations (Miura & Ogata, 2013 and references therein). The West-Japan and East-Japan geographic groups present homomorphic subtelocentric X and Y chromosomes, but they differ in morphology between the geographic groups, as the sex chromosomes of East-Japan specimens present a smaller short arm (Miura & Ogata, 2013 and references therein). Heteromorphic sex chromosomes are found in three other geographic groups, named the XY, ZW and Neo-ZW groups (Miura & Ogata, 2013 and references therein). Based on the analysis of mitochondrial DNA, allozymes, sex-linked genes and experimental reproductive crosses, two independent transitions from XX/XY to ZZ/ZW system could be inferred, one creating the ZW group and another giving rise to the Neo-ZW group. It is also very interesting to note that the metacentric X chromosome of the XY group and the metacentric W chromosome of both the ZW and neo-ZW groups are homeologous and probably originated from a pericentromeric inversion of the sex chromosome of the East-Japan group. The XY group and the ZW group were created by hybridization events between West-Japan and East Japan groups, whereas the neo-ZW group was derived from re-hybridization of the XY group with the West-Japan group (Miura & Ogata, 2013 and references therein).

Transition between female and male heterogamety and translocation of the sex-determining locus to an autosome (which gives rise to a new sex chromosome pair without changing the heterogametic sex) have been frequently invoked to explain the prevalence of homomorphic sex chromosomes in some groups, such as anurans (see discussion in Volff et al., 2007; Perrin, 2009; Evans et al., 2012). Such events of sex chromosome turnover interrupt the differentiation between the previous sex chromosomes as new sex chromosome systems are achieved.

In addition to sex chromosome turnover, the fountain of youth model (Perrin, 2009) could also explain the prevalence of homomorphic sex chromosomes in anurans. According to this model, occasional recombination events between X/Z and Y/W near the sex-specific region would avoid accumulation of deleterious mutations (Muller ratchet) and arrest the degeneration process. Besides the evidence of recent events of recombination between the X and Y chromosomes of *Hyla arborea* (Berset-Brändli et al., 2008), which was already considered by Perrin (2009), new evidence of occasional sex chromosome recombination has been reported to species closely related to *H. arborea* and to *H. meridionalis* and *H. suweonensis* (Stöck et al., 2011, 2013a; Guerrero et al., 2012; Dufresnes et al., 2015). Species from other groups, such as the *Bufotes viridis* group, also agree with fountain of youth model, since sex-linked markers are not grouped by gametolog in phylogenetic analysis, but by species (Stöck et al., 2013b).

Both models (sex determining transitions and fountain of youth) are supported by strong empirical evidence, which suggests that they may act together, as argued by Dufresnes et al., (2015) for the Palearctic tree frog lineages. In this hylid group, both male heterogamety (as in *Hyla arborea* and closely related species, *Hyla meridionalis*, and *Hyla japonica*[1]) and female (as in *Hyla suweonensis*) heterogamety are present, and evidence of occasional recombination events between sex chromosomes was found (Dufresnes et al., 2015 and references therein).

## 5.4. New Technologies in the Study of Sex Chromosomes in Anurans

Recently, next generation sequencing (NGS) has brought great progress to the study of sex chromosomes of several organisms. Different approaches using NGS have been employed and we summarize some of these as follows: (a) genome resequencing of heterogametic sex individuals and read mapping against a reference genome; (b) genome sequencing of males and females separately and comparing their *de novo* assemblies; (c) sequencing of isolated sex chromosomes obtained by laser capture microdissection or flow sorting; (d) transcriptome characterization of male and female gonads in several stages of embryonic development; and (e) genotyping by sequencing for genetic linkage map construction or identification of sex-linked markers.

Genome resequencing and read mapping against a reference genome have proven to be a reliable approach for investigating sex chromosomes, as in chicken (Chen et al., 2012) and in *Anopheles* mosquitoes (Hall et al., 2013). However, there are only three anuran genomes available to date: (a) *Xenopus tropicalis* (Hellsten et al., 2010); (b) *X. laevis* (Session et al.,

[1] In *Hyla arborea* and its closely related species *H. intermedia*, *H. molleri*, and *H. orientalis*, the linkage group 1 (LG1) is sex linked. LG1 is also sex linked in *H. meridionalis* and in *H. suweonensis*, which presents female heterogamety. In contrast, in *H. japonica*, which is inferred to be the sister species of *H. suweonensis* and presents male heterogamety, LG1 is not sex linked. This linkage group is not sex linked either in *H. sarda*, *H. savignyi*, or *H. felixarabica* (Dufresnes et al., 2015 and references therein). An evolutionary proposal for sex chromosome turnovers in this group was presented by Dufresnes et al. (2015), who also discussed the fact that LG1 was co-opted for sex determination several times independently.

2016) and (c) *Nanorana parkeri* (Sun et al., 2015). Since read mapping requires at minimum a good reference genome from a closely related species, currently this is not the best approach for anurans.

On the other hand, genotyping by sequencing (GBS; Elshire et al., 2011), which involves the sequencing of libraries of restriction fragments, has already shown interesting results in studies of sex chromosomes of frogs when data obtained from males and females are compared. Such an approach was satisfactorily used for the identification of sex-linked SNPs (single nucleotide polymorphisms) in *Xenopus borealis*, which helped Furman & Evans (2016) to propose that the DM-W gene is not the sex determination trigger in this species, although this gene is present in the other species of *Xenopus* closely related to *X. borealis*. In *Xenopus tropicalis* the data generated by restriction site-associated DNA sequencing (RAD-seq) together with the sanger sequencing of 65 amplicons supposedly linked to the sex-determining region suggested that the sex-specific region is small in this species, with large portion of its sex chromosomes corresponding to the pseudoautosomal region (Bewick et al., 2013). More recently, Brelsford et al. (2016) proposed a method for identifying sex chromosomes based on high-density linkage maps constructed from RAD-seq of individuals of a family without knowledge of offspring sexes. This method involves the identification of linkage groups (which correspond to the distinct chromosomes of the karyotype) and the construction of sex-specific maps based on the data obtained from the parents. The linkage group corresponding to the sex chromosome pair is inferred by comparing the number of informative markers per linkage group between the male- and female-specific maps. Because higher divergence is expected between the gametologs than between two chromosomes X (or Z), for the sex-specific linkage group a strong difference in the number of heterozygous markers between the male and female maps is expected, while for each autosome the number of informative markers is supposed to be similar between the two parental maps.

# 6. B Chromosomes in Anura: General Properties and Exceptions

The presence of supernumerary chromosomes, known as B chromosomes, is a characteristic found in the genomes of many eukaryote lineages, having been reported in all principal evolutionary branches, with many examples found in animals, plants, and fungi (for review, see Camacho, 2000). While we are still a long way from understanding the mechanisms involved in the origin, maintenance, and function of these chromosomes in genomes, they are characterized by three unifying features: (i) B chromosomes are dispensable to the principal genomic complement (A complement) and, when present, they may be limited to only a few populations of the species, and normally at low individual frequencies. Given this, new organisms in which these elements can be found are discovered sporadically. (ii) B chromosomes are inherited irregularly, that is, in a non-Mendelian way. (iii) During meiosis, B chromosomes do not pair with any chromosome of the A complement. In anurans, 22 species belonging to twelve different families (Table 3) are known to have B chromosomes, but this represents only a very small proportion of the total number of anuran species karyotyped up until now.

As in other eukaryotes, the B chromosomes of anurans are typically minor genomic elements in comparison with the A chromosomes (Nur & Nevo, 1969; Schmid, 1978; Schmid et al., 1987; Baldisera et al., 1993; Rosset et al., 2006; Medeiros et al., 2006; Schmid et al., 2010; Hernández-Guzmán et al., 2011; Ferro et al., 2016). There are exceptions, however, such as in *Physalaemus feioi* (Milani et al., 2010), *Oreobates discoidalis* (Schmid et al., 2010), and *Hymenochirus boettgeri* (Mezzasalma et al., 2015), in which the B chromosomes are at least as large as the chromosomes of the A complement.

**Table 3. Summary of the B chromosome systems described in Anura**

| Family/Species | B chromosomes | | | Chromosome markers | | Reference |
|---|---|---|---|---|---|---|
| | Number* | Morphology** | Type | Heterocromatin | rDNA sites | |
| Alytidae | | | | | | |
| *Discoglossus pictus* | 2 | S | t | n.e. | - | Schmid et al., 1987 |
| Hemiphractidae | | | | | | |
| *Gastrotheca espelitia* | 9 | D | m/t | C+ | + | Schmid et al., 2002; Schmid et al., 2012 |
| Hylidae | | | | | | |
| *Acris crepitans* | 5 | S | sm | n.e. | n.e. | Nur & Nevo, 1969 |
| *Bokermannohyla* | 1 | S | m | C+ (?) | - | Baldisera et al., 1993 |
| *luctuosa* | 3 | S | t | C- | - | Medeiros et al., 2006 |
| *Dendropsophus nanus* | 2 | D | m | q+pint/ q+p per | - | Cardoso et al., 1986; Gruber et al., 2007; |
| *Boana albopunctata* | 2 | S | t | n.e. | n.e. | Ferro et al., 2012 (as *Hypsiboas* |
| *Smilisca baudinii* | 2 | S | t | C+ | - | *albopunctatus*) |
| *Sphoenohyla dorsalis* | | | | | | Hernández-Guzmán et al., 2011 |
| | | | | | | Suaréz et al., 2013 |
| Hylodidae | | | | | | |
| *Megaelosia massarti* | 1 | D (2) | m/sm | C+/ q+pter | - | Rosa et al., 2003 |
| Leiopelmatidae | | | | | | |
| *Leiopelma hochstetteri* | 15 | D (3) | t | cen | - | Green et al., 1987; Green et al., 1993; Sharbel et al., 1998 |
| Ranidae | | | | | | |
| *Amolops* | 1 | S | t | cen | - | Wu & Zhao, 1985 |
| *liangshanensis* | 1 | S | t | cen | - | Kuramoto et al., 1989 |
| *Sanguirana everetti* | 4 | S | sm | qint | - | Schmid et al., 1978 |
| *Rana temporaria* | | | | | | |
| Scaphiopodiae | | | | | | |
| *Spea hammondii* | 1 | S | t | C+ | + | Green & Sessions, 1991 |

| Family/Species | B chromosomes | | | Chromosome markers | | Reference |
|---|---|---|---|---|---|---|
| | Number* | Morphology** | Type | Heterocromatin | rDNA sites | |
| Craugastoridae | | | | | | |
| Craugastor sp. B | 2 | S | m | C+ | - | Schmid et al., 2010 |
| *Oreobates discoidalis* | 1 | D (2) | t | n.e. | n.e. | Schmid et al., 2010 |
| *Oreobates barituensis* | 2 | D | t/st | cen | + | Ferro et al., 2016 |
| Eleutherodactylidae | | | | | | |
| *Eleutherodactylus gundlachi* | 10 | D | t | C+ | + | Schmid et al., 2010 |
| Leptodactylidae | | | | | | |
| *Physalaemus feioi* | 1 | S | st | qint | + (qper) | Milani et al., 2010; this work |
| Odontophrynidae | | | | | | |
| *Odontophrynus americanus* | 5 | S/D | st | qint | + (qper) | Rosset et al., 2006; Lanzone et al., 2008 |
| Pipidae | | | | | | |
| *Hymenochirus boettgeri* | 1 | S | m | C+ | - | Mezzasalma et al., 2015 |

* maximum number of B chromosomes per cell. ** (S) B chromosomes undifferentiated; (D) B chromosomes vary in size and morphology; [1] Polyploid species with inter-individual and inter-population variation;

Presence of NORs is indicated by a "+", while the absence of NORs is indicated by a "-".

q = long arm; p = short arm; C+ = completely heterochromatic; C- = absence of heterochromatin blocks; cen = centromere; int = interstitial position; per = pericentromeric position; ter = terminal position in the chromosome; n.e = not examine.

Rarely, as observed in *Megaelosia massarti* (Rosa et al., 2003), the B chromosome is the largest chromosome of the karyotype.

These supernumerary elements behave independently of the A complement and at a local population level, as indicated by the fact that different populations of the same species may present distinctive morphologies or numbers of B chromosomes as observed among populations of *Boana albopunctata* (Gruber et al., 2007; Ferro et al., 2012 – as *Hypsiboas albopunctatus*) and *Odontophrynus americanus* (Rosset et al., 2006; Lanzone et al., 2008). Distinct chromosome morphotypes may coexist within a given population, and may even be found in the same individual in some cases (Ferro et al., 2016). Individuals may also present different combinations of these elements (Green, 1988; Green et al., 1993; Schmid et al., 2010). Schmid et al. (2010) documented, for example, four types of B chromosome in *Eleutherodactylus gundlachi*, all of which were telocentric, with two large elements similar in size to pairs 6 and 12 of the A complement of this species, while the other two B chromosomes are much smaller than those of the A complement. These authors documented specimens of *E. gundlachi* with up to 10 additional copies of the B chromosomes, but, when present, the large B chromosomes were only found in one copy in cells, whereas different random combinations of the small B chromosomes were found.

Given all of the different potential combinations of B chromosomes, some authors have attempted to define the limit for the number of copies of these supernumerary elements that a cell would be able to support. This type of analysis is complex, given that, in extreme cases, as many as nine or ten B chromosomes may be found in the same genome, as in the cases of *Gastrotheca espelitia* (Schmid et al., 2002b, Schmid, 2012) and *Eleutherodactylus gundlachi* (Schmid et al., 2010), respectively, reaching the unexpected number of 15 in *Leiopelma hochstetteri* (Green, 1988; Green et al., 1993). What number of B chromosomes a eukaryote cell can support and how the chromatin would be rearranged to accommodate this additional genomic content are questions that demand further attention, and these animals provide potentially interesting research models. The fact that segregation is non-Mendelian results in a spectrum of combinations of B chromosomes in a population, ranging from individuals who have no supernumerary elements to those that have a number of copies (for example, see Green, 1988; Green et al., 1993).

The meiotic behavior of these supernumerary chromosomes is the key to their irregular segregation. By definition, B chromosomes do not pair with the chromosomes from complement A during meiotic prophase I, and the observation of univalent chromosomes in a species is one of the criteria used to confirm the presence of supernumerary elements in the genome. When more than one copy is present in the nucleus, these elements may pair with one another, presenting a bivalent (example in Ferro et al., 2016) or multivalent (example in Nur & Nevo, 1969) configuration. In other cases, however, these copies may remain univalent during meiosis I, as observed in male *Dendropsophus nanus*, which have three copies of euchromatic B chromosomes (Medeiros et al., 2006). The majority of B chromosomes, despite being unstable during meiosis, are stable during mitosis (example in Gruber et al., 2007; Ferro et al., 2016), although there are exceptions (as $B_d$ chromosome described by Ferro et al., 2016).

In practically all cases of B chromosomes in anurans, C-banding reveals elements either partially (see, for example, Schmid, 1978; Gruber et al., 2007; Milani et al., 2010; Ferro et al., 2012) or completely heterochromatic (Green & Sessions, 1991; Schmid et al., 2002b; Schmid et al., 2010; Suaréz et al., 2013; Mezzasalma et al., 2015), a characteristic also observed in the B chromosomes of other vertebrate groups. This would appear to be related to the evolutionary dynamics of these chromosomes because as dispensable elements in the genome, they are under weak selection pressure, which may lead to the accumulation of repetitive sequences (Camacho, 2000). In some organisms, the analysis of the heterochromatin content of B chromosomes has revealed the accumulation of transposons and retrotransposons, and the presence of different families of repetitive DNA. In anurans, unfortunately, the description of the heterochromatin content of the B chromosomes has been limited to conventional banding methods, while their nucleotide content has been analyzed only using base-specific fluorochromes (Gruber et al., 2007; Milani et al., 2010; Ferro et al., 2012; Mezzasalma et al., 2015; Ferro et al., 2016). However, small amounts of heterochromatin were already reported for supernumerary elements of *Dendropsophus nanus* (Medeiros et al., 2006) and in *Megaelosia massarti* (Rosa et al., 2003), for example, which makes these species interesting models for the study of the DNA content of these elements.

The fact that the B chromosomes accumulate large heterochromatic segments does not necessarily mean that they are transcriptionally inert. With the ongoing advances in genomic methods, this has been shown empirically in the B chromosomes of plants, grasshoppers, and fish. The B chromosomes of some anuran species contain rDNA sites which are transcriptionally active in the genome and can be detected through the silver Ag-NOR method (Green & Sessions, 1991; Milani et al., 2010; Schmid et al., 2010; Schmid et al., 2012; Ferro et al., 2016). Two copies of rDNA were detected by FISH in one of the B chromosome types of *Oreobates barituensis* (Ferro et al., 2016), while only one copy was detected by the Ag-NOR method, providing direct evidence of transcriptional activity prior to the last cell cycle.

The molecular composition of the B chromosomes is the principal source of evidence for the formulation of hypotheses on their origin. Three principal hypotheses have been proposed to account for the appearance of these chromosomes: (i) intraspecific origin, (ii) byproducts of the process of differentiation of the sex chromosomes, and (iii) interspecific origin, as vestiges of hybridization events. In the case of the intraspecific origin, the principal assumption is the sharing of DNA sequences with the A chromosomes genome. As mentioned above, the fact that B chromosomes do not recombine with the A genome means that these elements are inherited clonally, which means that we would expect young B chromosomes of intraspecific origin to present greater similarities with the A chromosomes. This hypothesis has been used to explain the origin of B chromosomes of a number of different groups of organisms. The supernumerary chromosome of *Sphaenohynchus dorsiae*, for example, contains a large block of telomeric repeats that occupy almost the whole of the B chromosome, which Suárez et al. (2013) found to be highly similar to the segment present in the homologs of pair 2 of *Sphaenohynchus lacteus*, which is absent from homeolog 2 of *S. dorsiae*. This evidence led the authors to suggest that the B chromosome of this species may have originated from this region of chromosome 2.

In the B chromosome system of *L. hochstetteri*, however, the comparison of the DNA sequences of the B chromosomes with those of the W chromosome revealed a high degree of similarity, which is consistent with the intraspecific origin of B elements from the diversification of the W sex chromosome of this species (Sharbel et al., 1998). Similar evidence of sex-chromosome derived B chromosomes has been found in *Characidium gomesi*, another species with a ZZ/ZW system (Pansonato-Alves et al., 2014).

The third hypothesis postulates that the B chromosomes are remnants of genomic hybridization events that have persisted in the genome. This hypothesis is normally the most difficult to validate, because it requires the confirmation of two fundamental aspects: (i) the sharing of specific DNA sequences between the B chromosome of one species and the A chromosomes from putative other species interbreed, and (ii) the occurrence of the two species within the same zone of sympatry. While a number of studies have attempted to prove the role of this mechanism in the origin of B chromosomes in anurans, no conclusive evidence of the process has been presented in any case.

While the origin of B chromosomes has been identified conclusively in some cases, most attempts to assess their homologies with the A chromosomes have failed, which may be a consequence of the independent evolution of B and A complements. In *Physalaemus feioi*, for example, the supernumerary chromosome has a large heterochromatin block on its long arm, which was exclusively detected by a probe derived from microdissected B chromosomes (Figure 5). No hybridization signal of this specific B probe was detected in the A complement, suggesting that some repetitive sequences are accumulated or exclusively present in this supernumerary chromosome. A similar situation has also been reported to the B chromosome of a Brazilian population of *Boana albopunctata*, which was entirely detected by a probe that did not hybridize to the A complement of this species (Gruber et al., 2014). In this case, the B probe detected small scattered regions on all chromosomes of another species, *Boana raniceps* (Gruber et al., 2014), which may correspond to sites of repetitive DNA sequences. Given the dynamic evolution of repetitive DNA (for review, see Feschotte & Pritham, 2007 and Plohl et al., 2008), the sharing of repetitive sequences should be analyzed with caution when assessing homology hypotheses. For this reason, the results observed by Gruber and colleagues should not be interpreted as a direct evidence of interspecific origin of the B chromosome of *B. albopunctata* (see discussion in Gruber et al., 2014).

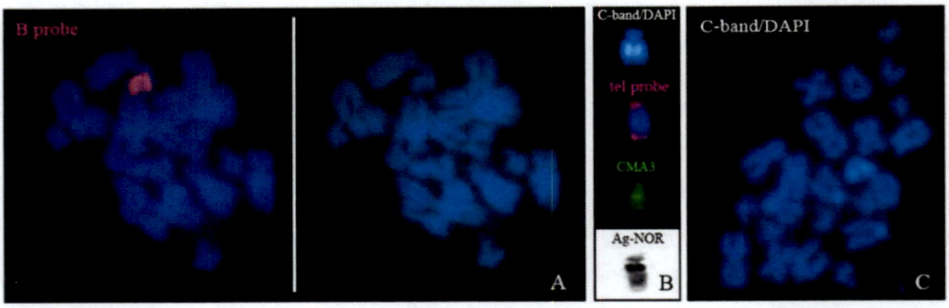


Figure 5. B chromosome of *Physalaemus feioi*. **A.** *In situ* hybridization of a probe derived from microdissected B chromosomes. On the right, DAPI image; on the left, merged image of DAPI and B probe. Note that no hybridization signal is seen in the A complement, whereas the long arm of the B chromosome is detected by the probe. **B**. B chromosome submitted to C-banding and DAPI staining (C-band+DAPI), *in situ* hybridized with telomeric probe (tel probe), stained with chromomycin A3 (CMA3), and silver stained (Ag-NOR). **C.** C-banded metaphase stained with DAPI. The chromosome preparations in **A** and **B** were obtained from the female specimen SMRP 371.28, whereas the metaphase shown in C was obtained from the male ZUEC 16249 (one of the specimens also analyzed by Milani et al., 2010). The B chromosome found in the specimen SMRP 371.28 differs from those described previously (Milani et al., 2010) by the presence of a pericentromeric NOR, although the specimens were all collected from Viçosa-MG, Brazil. ZUEC: Museum of Zoology, University of Campinas, Brazil. SMRP: Collection "Shirlei Maria Recco Pimentel", University of Campinas, Brazil.

Another interesting finding on the B chromosomes of anurans is the presence of these elements in the genomes of natural polyploid species. While B chromosomes are relatively common in polyploid plants (for review, see Palestis et al., 2004), up until now, *Odontophrynus americanus*, a South American frog, is the only polyploid vertebrate known to have B chromosomes (Rosset et al., 2006; Lanzone et al., 2008). Although these elements were rare in the *O. americanus* populations analyzed by Rosset et al. (2006) and Lanzone et al. (2008), this organism provides a potentially interesting model for the analysis of the origin and maintenance of these elements in the genome.

In conclusion, despite the general properties recognized among B chromosomes of the different anuran species, the evolutionary profiles of these enigmatic chromosomes are far from being completely understood.

## REFERENCES

Abramyan, J., Ezaz T., Graves, J. A. M. & Koopman, P. (2009). Z and W sex chromosomes in the cane toad (*Bufo marinus*). *Chromosome Res.*, *17*, 1015–1024.

Acosta, M. J., Marchal, J. A., Martínez, S. & Sánchez, A. (2007) Characterization of the satellite DNA Msat-160 from the species *Chionomys nivalis* (Rodentia, Arvicolinae). *Genetica*, *130*, 43

Aguiar-Jr, O., Bacci, Jr. M., Lima, A. P., Rossa-Feres, D. C., Haddad, C. F. B. & Recco-Pimentel, S. M. (2007). Phylogenetic relationships of *Pseudis* and *Lysapsus* (Anura, Hylidae, Hylinae) inferred from mitochondrial and nuclear gene sequences. *Cladistics*, *23*, 455-463.

Altinordu, F., Peruzzi, L., Yu, Y. & He, X. (2016). A tool for the analysis of chromosomes: Karyotypes. *Taxon*, *65*, 586-592.

Amaro, R. C., Rodrigues, M. T., Yonenaga-Yassuda, Y. & Carnaval, A. C. (2012). Demographic processes in the montane Atlantic rainforest: Molecular and cytogenetic evidence from the endemic frog *Proceratophrys boiei*. *Mol. Phylogenet. Evol.*, *62*, 880–888.

Amaro-Ghilardi, R. C., Silva, M. J. J., Rodrigues, M. T. & Yonenaga-Yassuda, Y. (2007) Chromosomal studies in four species of genus *Chaunus* (Bufonidae, Anura): localization of telomeric and ribosomal sequences after fluorescence *in situ* hybridization (FISH). *Genetica*, *134*, 159-168.

Ananias, F., Modesto, A. D. S., Mendes, S. C. & Napoli, M. F. (2007). Unusual primitive heteromorphic ZZ/ZW sex chromosomes in *Proceratophrys boiei* (Anura, Cycloramphidae, Alsodinae), with description of C-Band interpopulational polymorphism. *Hereditas*, *144*, 206–212.

Andrade, F. S. & Carvalho, T. R. (2013). A new species of *Pseudopaludicola* Miranda-Ribeiro (Leiuperinae: Leptodactylidae: Anura) from the Cerrado of southeastern Brazil. *Zootaxa*, *3608*, 389–397.

Andrade, F. S., Magalhães, F. M., Nunes-de-Almeida, C. H. L., Veiga-Menoncello, A. C. P., Santana, D. J., Garda, A. A., Loebmann, D., Recco-Pimentel, S. M., Giaretta, A. A. & Toledo, L. F. (2016). A new species of long-legged *Pseudopaludicola* from northeastern Brazil (Anura, Leptodactylidae, Leiuperinae). *Salamandra*, *52*, 107-124.

Baldissera, F. A., Deoliveira, P. S. L. & Kasahara, S. (1993). Cytogenetics of 4 Brazilian *Hyla* species (Amphibia- Anura) and description of a case with a supernumerary chromosome. *Rev. Bras. Genet.*, *16*, 335-345.

Bergero, R. & Charlesworth, D. (2009). The evolution of restricted recombination in sex chromosomes. *Trends Ecol. Evol.*, *24*, 94-102.

Berneck, B. V. M., Haddad, C. F. B., Lyra, M. K., Cruz, C. A. G. & Faivovich, J. (2016). The green clade grows: A phylogenetic analysis of *Aplastodiscus* (Anura, Hylidae). *Mol. Phylogen. Evol.*, *97*, 213-223.

Berset-Brändli, L., Jaquiéry, J. & Perrin, N. (2007). Recombination is suppressed and variability reduced in a nascent Y chromosome. *J. Evol. Biol.*, *20*, 1182-1188.

Berset-Brändli, L., Jaquiéry, J., Broquet, T., Ulrich, Y. & Perrin, N. (2008). Extreme heterochiasmy and nascent sex chromosomes in European tree frogs. *Proc. R. Soc. B*, *275*, 1577-1585.

Berta, P., Hawkins, J. R., Sinclair, A. H., Taylor, A., Griffiths, B. L., Goodfellow, P. N. & Fellous, M. (1990). Genetic evidence equating SRY and the testis-determining factor. *Nature*, *348*, 448-450.

Bewick, A. J., Anderson, D. W. & Evans, B. J. (2011). Evolution of the closely related, sex-related genes DM-W and DMRT1 in African clawed frogs (*Xenopus*). *Evolution*, *65*, 698-712.

Bewick, A. J., Chain, F. J. J., Zimmerman, L. B., Sesay, A., Gilchrist, M. J., Owens, N. D. L., Seifertova, E., Krylov, V., Macha, J., Tlapakova, T., Kubickova, S., Cernohorska, H., Zarsky, V. & Evans, B. J. (2013). A large pseudoautosomal region on the sex chromosomes of the frog *Silurana tropicalis*. *Genome Biol. Evol.*, *5*, 1087-1098.

Blotto, B. L., Nuñez, J. J., Basso, N. G., Úbeda, C. A., Wheeler, W. C. & Faivovich, J. (2013). Phylogenetic relationships of a Patagonian frog radiation, the *Alsodes* + *Eupsophus* clade (Anura: Alsodidae), with comments on the supposed para-phyly of *Eupsophus*. *Cladistics*, *29*, 113-131.

Borgart, J. P. (1991). The influence of life history on karyotipic evolution in frogs. In *Amphibian cytogenetics and evolution* (ed. D. M. Green & S.K. Sessions), pp. 233-257. Academic Press, San Siego, USA.

Brelsford, A., Dufresnes, C. & Perrin, N. (2016). High-density sex-specific linkage maps of a European tree frog (*Hyla arborea*) identify the sex chromosome without information on offspring sex. *Heredity*, *116*, 177-181.

Busin, C. S., Viciprova, G. & Recco-Pimentel, S. M. (2011). Chromosomal rearrangements as the source of variation in the number of chromosomes in *Pseudis* (Amphibia, Anura). *Genetica*, *110*, 131-141.

Busin, C. S., Andrade, G. V., Bertoldo, J., Del-Grande, M. L., Uetanabaro, M. & Recco-Pimentel, S. M. (2008). Cytogenetic analysis of four species of *Pseudis* (Anura, Hylidae), with the description of ZZ/ZW sex chromosomes in *P. tocantins*. *Genetica*, *133*, 119-127.

Camacho, J. P. M., Sharbel, T. F. & Beukeboom, L. W. (2000) B chromosome evolution. *Philos. T. R. Soc. B*, *355*, 163-178.

Caminer, M. A., Mila,´ B., Jansen, M., Fouquet, A., Venegas, P. J., Chávez, G., Lougheed, S. C. & Ron, S. R. (2017). Systematics of the *Dendropsophus leucophyllatus* species complex (Anura: Hylidae): Cryptic diversity and the description of two new species. *PLoS ONE*, *12*(3), e0171785.

Cardozo, D. E., Leme, D. M., Bortoleto, J. F., Catroli, G. F., Baldo, D., Faivovich, J., Kolenc, F., Silva, A. P. Z., Borteiro, C., Haddad, C. F. B. & Kasahara, S. (2011). Karyotypic data on 28 species of *Scinax* (Amphibia: Anura: Hylidae): diversity and informative variation. *Copeia*, *2*, 251-263.

Cazaux, B., Catalan, J., Justy, F., Escudé, C., Desmarais, E. & Britton-Davidian, J. (2013). Evolution of the structure and composition of house mouse satellite DNA sequences in the subgenus *Mus* (Rodentia: Muridea): a cytogenomic approach. *Chromosoma*, *122*, 209-220.

Charlesworth, D., Charlesworth, B. & Marais, G. (2005). Steps in the evolution of heteromorphic sex chromosomes. *Heredity*, *95*, 118-128.

Chen, J. M., Zhou, W. W., Poyarkov, Jr. N. A., Stuart, B. L., Brown, R. M., Lathrop, A., Wang, Y. Y., Yuan, Z. Y., Jiang, K., Hou, M., Chen, H. M., Suwannapoom, C., Nguyen, S. N., Duong, T. V., Papenfuss, T. J., Murphy, R. W., Zhang, Y. P. & Che, J. (2017). A novel multilocus phylogenetic estimation reveals unrecognized diversity in Asian horned toads, genus *Megophrys* sensu lato (Anura: Megophryidae). *Mol. Phylogenet. Evol.*, *106*, 28-43.

Cortez, D., Marin, R., Toledo-Flores, D., Froidevaux, L., Liechti, A., Waters, P. D., et al. (2014). Origins and functional evolution of Y chromosomes across mammals. *Nature*, *508*, 488-493.

Cuevas, C. C. & Formas, J. R. (1996). Heteromorphic sex chromosomes in *Eupsophus insularis* (Amphibia: Anura: Leptodactylidae). *Chromosome Res.*, *4*, 467-70.

Cuevas, C. C. (2008). A new species of the genus *Alsodes* (Anura: Neobatrachia) from the *Nothofagus* forest, Coastal Range, Southern Chile, identified by its karyotype. *Zootaxa*, *1771*, 43-53.

Cunningham, M. J. & Cherry, M. I. (2004). Molecular systematics of African 20-chromosome toads (Anura: Bufonidae). *Mol. Phylogenet. Evol.*, *32*, 671-685.

De Sá, R. O., Grant, T., Camargo, A., Heyer, W. R., Ponssa, M. L. & Stanley, E. (2014). Systematics of the Neotropical genus *Leptodactylus* Fitzinger, 1826 (Anura: Leptodactylidae): phylogeny, the relevance of non-molecular evidence, and species accounts. *S. Am. J. Herpetol.*, *9*, S1–S100.

Diniz, D., Laudicina, A., Cioffi, M. B. & Bertollo, L. A. C. (2008). Microdissection and whole chromosome painting. Improving sex chromosome analysis in *Triportheus* (Teleostei, Characiformes). *Cytogenet. Genome Res.*, *122*, 163-168.

Dover, G. (1982). Molecular drive: a cohesive mode of species evolution. *Nature*, *299*, 111-117.

Dover, G. A. (1986). Molecular drive in a multigene families: how biological novelties arise, spread and are assimilates. *Trends in Genet.*, *2*, 159-165.

Duarte, T. C., Veiga-Menoncello, A. C. P., Lima, J. F. R., Strüssmann, C., Del-Grande, M. L., Giaretta, A. A., Pereira, E. G., Rossa-Feres, D. C. & Recco-Pimentel, S. M. (2010). Chromosome analysis in *Pseudopaludicola* (Anura, Leiuperidae), with description of sex chromosomes XX/XY in *P. saltica*. *Hereditas*, *147*, 43-52.

Duellman, W. E., Marion, A. B. & Hedges, S. B. (2016). Phylogenetics, classification, and biogeography of the treefrogs (Amphibia: Anura: Arboranae). *Zootaxa.*, *4104*, 001-109.

Dufresnes, C., Borzée, A., Horn, A., Stöck, M., Ostini, M., Sermier, R., Wassef, J., Litvinchuck, S. N., Kosch, T. A., Waldman, B., Jang, Y., Brelsford, A. & Perrin, N. (2015). Sex-chromosome homomorphy in Palearctic tree frogs results from both turnovers and X - Y recombination. *Mol. Biol. Evol.*, *32*, 2328-2337.

Elder, J. F. & Turner, B. J. (1995). Concerted evolution of repetitive DNA sequences in eukaryotes. *Q. Rev. Biol.*, *70*, 297-320.

Elshire, R. J., Glaubitz, J. C., Sun, Q., Poland, J. A., Kawamoto, K., Buckler, E. S. & Mitchell, S. E. (2011). A robust, simple genotyping-by-sequencing (GBS) approach for high diversity species. *PLoS ONE*, *6*(5), e19379.

Evans, B. J., Kelley, D. B., Tinsley, R. C., Melnick, D. J. & Cannatella, D. C. (2004). A mitochondrial DNA phylogeny of clawed frogs: phylogeography on sub-Saharan Africa and implications for polyploid evolution. *Mol. Phylogenet. Evol.*, *33*, 197-213.

Evans, B. J. (2007). Ancestry influences the fate of duplicated genes millions of years after polyploidization of clawed frogs (*Xenopus*). *Genetics*, *176*, 1119-1130.

Evans, B. J. (2008). Genome evolution and speciation genetics of clawed frogs (*Xenopus* and *Silurana*). *Front. Biosci.*, *13*, 4687-4706.

Evans, B. J., Pyron, R. A. & Wiens, J. J. (2012) Polyploidization and Sex Chromosome Evolution in Amphibians. In: P. S. Soltis & D. E. Soltis (Eds.), *Polyploidy and Genome Evolution*. Springer-Verlag, Berlin, Heidelberg, pp. 385-410.

Evans, B. J., Carter, T. F., Greenbaum, E., Gvoždík, V., Kelley, D. B., McLaughlin, P. J., Pauwels, O. S. G., Portik, D. M., Stanley, E. L., Tinsley, R. C., Tobias, M. L. & Blackburn, D. C. (2015). Genetics, morphology, advertisement calls, and historical records distinguish six new polyploid species of African clawed frog (*Xenopus*, Pipidae) from West and Central Africa. *PLoS ONE*, *10*(12), e0142823.

Faivovich, J. (2002). A cladistic analysis of *Scinax* (Anura: Hylidae). *Cladistics*, *18*, 367-393.

Faivovich, J., Haddad, C. F. B., Garcia, P. C. A., Frost, D. R., Campbell, J. R. & Wheeler, W. C. (2005). Systematic review of the frog family Hylidae, with special reference to Hylinae: phylogenetic analysis and taxonomic revision. *B. Am. Mus. Nat. Hist.*, *294*, 1-240.

Faivovich, J., Ferraro, D. P., Basso, N. G., Haddad, C. F. B., Rodrigues, M. T.; Wheeler, W. C. & Lavilla, E. O. (2012). A phylogenetic analysis of *Pleurodema* (Anura: Leptodactylidae: Leiuperinae) based on mitochondrial and nuclear gene sequences, with comments on the evolution of anuran foam nests. *Cladistics*, *28*, 460-482.

Fávero, E. R., Veiga-Menoncello, A. C. P., Rossa-Feres, D. C., Strüssmann, C., Giaretta, A. A., Andrade, G. V., Colombo, P. & Recco-Pimentel, S. M. (2011). Intrageneric karyotypic variation in *Pseudopaludicola* (Anura: Leiuperidae) and its taxonomic relatedness. *Zool. Stud.*, *50*, 826-836.

Feng, Y. J., Blackburn, D. C., Liang, D., Hillis, D. M., Wake, D. B., Cannatella, D. C. & Zhang, P. (2017). Phylogenomics reveals rapid, simultaneous diversification of three major clades of Gondwanan frogs at the Cretaceous-Paleogene boundary. *P. Natl. Acad. Sci. USA*, *114*, E5864-E5870.

Ferro, M. J., Taffarel, A., Cardozo, D., Grosso, J., Puig, M. O., Suárez, P., Akmentins, M. S. & Baldo, D. (2016). Cytogenetic characterization and B chromosome diversity in direct-developing frogs of the genus *Oreobates* (Brachycephaloidea, Craugastoridae). *Comp. Cytogenet.*, *10*, 141-156.

Feschotte, C. & Pritham, E. J. (2007). DNA transposons and the evolution of eukaryotic genomes. *Annu. Rev. Genet.*, *41*, 331-368.

Fouquet, A., Gilles, A., Vences, M., Marty, C., Blanc, M. & Gemmell, N. J. (2007). Underestimation of species richness in Neotropical frogs revealed by rDNA analyses. *PLoS ONE*, *2*(10), e1109.

Frost, D. R. *Amphibian species of the world: an online reference*. Version 6.0 (Date of access: 15/nov/2017). Available from: URL: http:// research.amnh.org/herpetology/ amphibia/index.html.

Funk, W. C., Caminer, M. & Ron, S. R. (2012). High levels of cryptic species diversity uncovered in Amazonian frogs. *Proc. R. Soc. B.*, *279*, 1806-1814.

Furman, B. L. & Evans, B. J. (2016). Sequential turnovers of sex chromosomes in African clawed frogs (*Xenopus*) suggest some genomic regions are good at sex determination. *G3-Genes Genom. Genet.*, *6*, 3625-3633.

Gatto, K. P., Busin, C. S. & Lourenço, L. B. (2016). Unraveling the sex chromosome heteromorphism of the paradoxical frog *Pseudis tocantins*. *PLoS ONE*, *11*, e0156176.

Giam, X., Scheffers, B. R., Sodhi, N. S., Wilcove, D. S., Ceballos, G. & Ehrlich, P. R. (2013). Reservoirs of richness: least disturbed tropical forests are centers of undescribed species diversity. *Proc. R. Soc. B*, *279*, 67-76.

Goodfellow, P. N. & Lovell-Badge, R. (1993). SRY and sex determination in mammals. *Annu. Rev. Genet.*, *27*, 71-92.

Grant, T., Frost, D. R., Caldwell, J. P., Galiardo, R., Haddad, C. F. B., Kok, P. J. R., Means, D. B., Noonan, B. P., Schargel, W. E. & Wheeler, W. W. (2006). Phylogenetic systematic of dart-poison frogs and their relatives (Amphibia, Athesphatanura, Dendrobatidae). *B. Am. Mus. Nat. Hist.*, *299*, 1-262.

Graves, J. A. M. (2006). Sex chromosome specialization and degeneration in mammals. *Cell*, *124*, 901-914.

Graves, J. A. M. (2008). Weird animal genomes and the evolution of vertebrate sex and sex chromosomes. *Annu. Rev. Genet.*, *42*, 565-586.

Graves, J. A. M. & Peichel, C. L. (2010). Are homologies in vertebrate sex determination due to shared ancestry or to limited options? *Genome Biol.*, *11*, 205.

Green, D. M., Kezer, J. & Nussbaum, R. A. (1987). Supernumerary chromosome variation and heterochromatin distribution in the endemic New Zealand frog, *Leiopelma hochstetteri*. *Chromosoma*, *95*, 339-344.

Green, D. M. (1988). Cytogenetics of the endemic New Zealand frog, *Leiopelma hochstetteri*: extraordinary supernumerary chromosome variation and a unique sex-chromosome system. *Chromosoma*, *97*, 55-70.

Green, D. M. & Sessions, S. K. (1991). (eds): *Amphibian Cytogenetics and Evolution*. Academic Press, San Diego.

Green, D. M., Zeyl, C. W. & Sharbel, T. F. (1993). The evolution of hypervariable sex and supernumerary chromosomes in the relict New Zealand frog, *Leiopelma hochstetteri*. *J. Evol. Biol.*, *6*, 417-441.

Gruber, S. L., Haddad, C. F. B. & Kasahara, S. (2007). Chromosome banding in three species of *Hypsiboas* (Hylidae, Hylinae), with special reference to a new case of a B-chromosome in anuran frogs and to the reduction of the diploid number of 2n = 24 to 2n = 22 in the genus. *Genetica*, *130*, 281-291.

Gruber, S. L., Zina, J., Narimatsu, H., Haddad, C. F. B. & Kasahara, S. (2012a). Comparative karyotype analysis and chromosome evolution in the genus *Aplastodiscus* (Cophomantini, Hylinae, Hylidae). *BMC Genet.*, *13*, 28.

Gruber, S., Haddad, C. F. B. & Kasahara, S. (2012b). Karyotype analysis of seven species of the tribe Lophiohylini (Hylinae, Hylidae, Anura), with conventional and molecular cytogenetic techniques. *Comp. Cytogenet.*, *6*, 409-423.

Gruber, S. L., Diniz, D., Sobrinho-Scudeler, P. E., Foresti, F., Haddad, C. F. B. & Kasahara, S. (2014). Possible interspecific origin of the B chromosome of *Hypsiboas albopunctatus* (Spix, 1824) (Anura, Hylidae), revealed by microdissection, chromosome painting, and reverse hybridization. *Comp. Cytogenet.*, *8*, 185-197.

Guarnizo, C. E., Escallón, C., Cannatella, D. C. & Amésquita, A. (2012). Congruence between acoustic traits and genealogical history reveals a new species of *Dendropsophus* (Anura: Hylidae) in the high Andes of Colombia. *Herpetologica*, *68*, 523-540.

Guerrero, R. F., Kirkpatrick, M. & Perrin, N. (2012). Cryptic recombination in the ever-young sex chromosomes of Hylid frogs. *J. Evol. Biol.*, *25*, 1947-1954.

Hall, A. B., Qi, Y., Timoshevskiy, V., Sharakhova, M. V., Sharakhov, I. V. & Tu, Z. (2013). Six novel Y chromosome genes in *Anopheles* mosquitoes discovered by independently sequencing males and females. *BMC Genomics*, *14*, 273.

Hassanane, M. S., Chaudhary, R. & Chowdhary, B. P. (1998). Microdissected bovine X chromosome segment delineates homoeologous chromosomal regions in sheep, goat and buffalo. *Chromosome Res.*, *6*, 213-217.

Hellsten, U., Harland, R. M., Gilchrist, M. J., Hendrix, D., Jurka, J., Kapitonov, V., Ovcharenko, I., Putnam, N. H., Shu, S., Taher, L., Blitz, I. L., Blumberg, B., Dichmann, D. S., Dubchak, I., Amaya, E., Detter, J. C., Fletcher, R., Gerhard, D. S., Goodstein, D., Graves, T., Grigoriev, I. V., Grimwood, J., Kawashima, T., Lindquist, E., Lucas, S. M., Mead, P. E., Mitros, T., Ogino, H., Ohta, Y., Poliakov, A. V., Pollet, N., Robert, J., Salamov, A., Sater, A. K., Schmutz, J., Terry, A., Vize, P. D., Warren, W. C., Wells, D., Wills, A., Wilson, R. K., Zimmerman, L. B., Zorn, A. M., Grainer, R., Grammer, T., Khokha, M. K., Richardson, P. M. & Rokshar, D. S. (2010). The genome of the Western clawed frog *Xenopus tropicalis*. *Science*, *328*, 633-636.

Henikoff, S., Ahmad, K. & Malik, H. S. (2001). The centromere paradox: stable inheritance with rapidly evolving DNA. *Science*, *293*, 1098-1102.

Henning, F., Moysés, C. B., Calcagnotto, D., Meyer, A. & de Almeida-Toledo, L. F. (2011). Independent fusions and recent origins of sex chromosomes in the evolution and diversification of glass knife fishes (*Eigenmannia*). *Heredity*, *106*, 391-400.

Hernández-Guzmán, J., Arias-Rodriguez, L. & Indy, J. R. (2011). Los cromosomas meióticos de la rana arborícola *Smilisca baudinii* (Anura: Hylidae). *Rev. Biol. Trop.*, *59*, 355-362. [Meiotic chromosomes of the tree frog *Smilisca baudinii* (Anura: Hylidae)].

Hillis, D. M. & Green, D. M. (1990). Evolutionary phylogenetic changes of heterogametic history of amphibians. *J. Evol. Biol.*, *3*, 49-64.

Holloway, A. K., Cannatella, D. C., Gerhardt, H. C. & Hillis, D. M. (2006). Polyploids with different origins and ancestors form a single sexual polyploid species. *Am. Nat.*, *167*, E88-E101.

Iturra, P. & Veloso, A. (1989) Further evidence for early sex chromosome differentiation of anuran species. *Genetica*, *78*, 25-31.

Karamysheva, T. V., Andreenkova, O. V., Bochkaerev, M. N., Borissov, Y. M., Bogdanchikova, N, Borodin, P. M. & Rubtsov, N. B. (2002). B chromosomes of Korean field mouse Apodemus peninsulae (Rodentia, Murinae) analysed by microdissection and FISH. *Cytogenet. Genome Res.*, *96*, 154-160.

Kasahara, S., Silva, A. P. & Haddad, C. F. B. (1996). Chromosome banding in three species of Brazilian toads (Amphibia - Bufonoidae). *Braz J Gen*, *19*, 237-242.

Kawagoshi, T., Nishida, C. & Matsuda, Y. (2012). The origin and differentiation process of X and Y chromosomes of the black marsh turtle (*Siebenrockiella crassicollis*, Geoemydidae, Testudines). *Chromosome Res.*, *20*, 95-110.

Kirov, I., Khrustaleva, L., Van Laere, K., Soloviev, A., Meeus, S., Romanov, D. & Fesenko, I. (2017). DRAWID: user-friendly java software for chromosome measurements and ideogram drawing. *Comp. Cytogenet.*, *11*, 747-757.

Kohler, J., Vieites, D. R., Bonett, R. M., Garcia, F. H., Glaw, F., Steinke, D. & Vences, M. (2005). New amphibians and global conservation: A boost in species discoveries in a highly endangered vertebrate group. *Bioscience*, *55*, 693-696.

Krylov, V., Kubickova, S., Rubes, J., Macha, J., Tlapakova, T., Seifertova, E. & Sebkova N. (2010). Preparation of *Xenopus tropicalis* whole chromosome painting probes using laser microdissection and reconstruction of *X. laevis* tetraploid karyotype by Zoo-FISH. *Chromosomes Res.*, *1*, 431-439.

Kuramoto, M. (1989). Karyological studies on some Philippine frogs. In: M. Matsui, T. Kikida, R. C. Goris (Eds.), *Current Herpetology in East Asia.*, (pp 115–121) Kyoto, Herpetological Society of Japan.

Lanzone, C., Baldo, D. & Rosset, S. D. (2008). Meiotic differentiation in two allopatric population groups of the tetraploid frog *Odontophrynus americanus* from Argentina. *Herpetol. J.*, *18*, 213-222.

López-Flores, I. & Garrido-Ramos, M. A. (2012). The repetitive DNA content of eukaryotic genomes. In: Garrido-Ramos, MA. *Repetitive DNA*. Basel, 1-28.

Lourenço, L. B., Bacci, Jr. M., Martins, V. G., Recco-Pimentel, S. M. & Haddad, C. F. B. (2008). Molecular phylogeny and karyotype differentiation in *Paratelmatobius* and *Scythrophrys* (Anura, Leptodactylidae). *Genetica*, *132*, 255-266.

Lourenço, L. B., Targueta, C. P., Baldo, D., Nascimento, J., Garcia, P. C. A., Andrade, G. V., Haddad, C. F. B. & Recco-Pimentel, S. M. (2015). Phylogeny of frogs from the genus *Physalaemus* (Anura, Leptodactylidae) inferred from mitochondrial and nuclear gene sequences. *Mol. Phylogenet. Evol.*, *92*, 204-216.

Lyra, M. L., Haddad, C. F. & Azeredo-Espin, A. M. L. (2016). Meeting the challenge of DNA barcoding Neotropical amphibians: Polymerase chain reaction optimization and new COI primers. *Mol. Ecol. Resour.*, *17*, 966-980.

Mahony, M. J. (1991). Heteromorphic sex chromosomes in the Australian frog *Crinia bilingua* (Anura, Myobatrachidae). *Genome*, *34*(3), 334-337.

Marchal, J. A., Acosta, M. J., Nietzel, H., Sperling, K., Bullejos, M., Díaz-de-la-Guardia, R. & Sánchez, A. (2004). X chromosome painting in

*Microtus*: Origin and evolution of the giant sex chromosomes. *Chromosome Res.*, *12*, 767-776.

Matsuda, M., Shinomiya, A., Kinoshita, M., Suzuki, A., Kobayashi, T., Paul-Prasanth, B., Lau, E., Hamaguchi, S., Sakaizumi, M. & Nagahama, Y. (2007). DMY gene induces male development in genetically female (XX) medaka fish. *P. Natl. Acad. Sci. USA*, *104*, 3865-3870.

Matsubara, K., Uno, Y., Srikulnath, K., Seki, R., Nishida, C. & Matsuda, Y. (2015). Molecular cloning and characterization of satellite DNa sequences from constitutive heterochromatin of the habu snake (*Potobothrops flavoviridis*, Viperidae) and the Burmese phyton (*Phyton bivittatus*, Pythonidae). *Chromosoma*, *124*, 529-539.

Mawaribuchi, S., Takahashi, S., Wada, M., Uno, Y., Matsuda, Y., Kondo, M., Fukui, A., Takamatsu, N., Taira, M. & Ito, M. (2017). Sex chromosome differentiation and the W- and Z-specific loci in *Xenopus laevis*. *Dev. Biol.*, *426*, 393-400.

Medeiros, L. R., Rossa-Feres, D. C. & Recco-Pimentel, S. M. (2003). Chromosomal differentiation of *Hyla nana* and *Hyla sanborni* (Anura, Hylidae) with a description of NOR polymorphism in *H. nana*. *J. Hered.*, *94*, 149-154.

Medeiros, L. R., Rossa-Feres, D. C., Jim, J. & Recco-Pimentel, S. M. (2006). B-chromosomes in two Brazilian populations of *Dendropsophus nanus* (Anura, Hylidae). *Genet. Mol. Biol.*, *29*, 257-262.

Medeiros, L. R., Lourenço, L. B., Rossa-Feres, D. C., Lima, A. P., Andrade, G. V., Giaretta A., Egito, G. T. B. T. & Recco-Pimentel, S. M. (2013). Comparative cytogenetic analysis of some species of the *Dendropsophus microcephalus* group (Anura, Hylidae) in the light of phylogenetic inferences. *BMC Genetics*, *14*, 59.

Mestrovic, N., Plohl, M., Mravinac, B. & Ugarkovic, D. (1998). Evolution of satellite DNAs from the genus *Palorus*– experimental evidence for the 'library' hypothesis. *Mol. Biol. Evol.*, *15*, 1062-1068.

Mezzasalma, M., Gaetano, F. G., Petraccioli, O. A. & Guarino, F. M. (2005). Chromosome analyses of *Pseudhymenochirus merlini* and

*Hymenochirus boettgeri* provide new insights into the chromosome evolution in the anuran family Pipidae. *Zool. Anz.*, *258*, 47-53.

Milani, M., Cassini, C. S., Recco-Pimentel, S. M. & Lourenço, L. B. (2011). Karyotypic data detect interpopulational variation in *Physalaemus olfersii* and the first case of supernumerary chromosome in the genus. *Animal Biol. J.*, *2*, 21–28.7

Mirzaghaderi, G. & Marzangi, K. (2015). Ideo Kar: an ideogram constructing ande karyotype analyzing software. *Caryologia*, *68*, 31-35.

Miura, I. & Ogata, M. (2013). Change of heterogametic sex from male to female: Why so easy in the frog? *Chrom. Sci.*, *16*, 3-9.

Muller, H. J. (1964). The relation of recombination to mutational advance. *Mutat. Res.*, *1*, 2-9.

Nanda, I., Zend-Ajusch, E., Shan, Z., Grützner, F., Schartl, M., Burt, D. W., Koehler, M., Fowler, V. M., Goodwin, G., Schneider, W. J., Mizuno, S., Dechant, G., Haaf, T. & Schmid, M. (2000). Conserved synteny between the chicken Z sex chromosome and human chromosome 9 includes the male regulatory gene DMRT1: a comparative (re)view on avian sex determination. *Cytogenet. Cell Genet.*, *89*, 67-78.

Nanda, I., Schlegelmilch, K., Haaf, T., Schartl, M. & Schmid, M. (2008). Synteny conservation of the Z chromosome in 14 avian species (11 families) supports a role for Z dosage in avian sex determination. *Cytogenet. Genome Res.*, *122*, 150-156.

Nascimento, J., Quinderé, Y. R. S. D., Recco-Pimentel, S. M., Lima, J. R. F. & Lourenço, L. B. (2010). Heteromorphic Z and W sex chromosomes in *Physalaemus ephippifer* (Steindachner, 1864) (Anura, Leiuperidae). *Genetica*, *138*, 1127-1132.

Nur, U. & Nevo, E. (1969). Supernumerary chromosomes in the cricket frog *Acris crepitans*. *Caryologia*, *22*, 97-102.

Odierna, G., Aprea, G., Capriglione, T., Castellano, S. & Balletto, E. (2007). Cytological evidence for population-specific sex chromosome heteromorphism in Palaearctic green toads (Amphibia, Anura). *J. Biosci.*, *32*, 763-768

Ohno, S. (1967). *Sex chromosomes and sex-linked genes*. New York, Springer-Verlag.

Orrico, V. G. D., Nunes, I., Mattedi, C., Fouquet, A., Lemos, A. W., Rivera-Correa, M., Lyra, M., Loebmann, D., Pimenta, B. V. S., Caramaschi, U., Rodrigues, M. T. & Haddad, C. F. B. (2017). Integrative taxonomy supports the existence of two distinct species within *Hypsiboas crepitans* (Anura: Hylidae). *Salamandra*, *53*, 99-113.

Orrico, V. G. D., Duellman, W. E., Souza, M. B. & Haddad, C. F. B. (2013). The taxonomic status of *Dendropsophus allenorum* and *D. timbeba* (Anura: Hylidae). *J. Herpetol.*, *46*, 615-618.

Ohta, S., Sumida, M. & Nishioka, M. (1999). Sex-determining mechanism in *Buergeria buergeri* (Anura, Rhacophoridae). III. Does the ZZW triploid frog become female or male? *J. Exp. Zool.*, *283*, 295–306.

Padial, J. M., Grant, T. & Frost, D. R. (2014). Molecular Systematics of terraranas (Anura: Brachycephaloidea) with an assessment of the effects of alignment and optimality criteria. *Zootaxa*, *3825*, 1-132.

Palestis, B. G., Trivers, R., Burt, A. & Jones, R. N. (2004). The distribution of B chromosomes across species. *Cytogenet. Genome Res.*, *106*, 151-158.

Pansonato, A., Veiga-Menoncello, A. C. P., Mudrek, J. R., Martin, J., Recco-Pimentel, S. M. & Strussmann, C. (2016). Two New Species of *Pseudopaludicola* (Anura: Leptodactylidae: Leiuperinae) from Eastern Bolivia and Western Brazil. *Herpetologica*, *72*, 235-255.

Pansonato-Alves, J. C., Serrano, E. A., Utsunomia, R., Camacho, J. P. M., Costa-Silva, G. J., Vicari, M. R., Artoni, R. F., Oliveira, C. & Foresti, F. (2014). Single origin of sex chromosomes and multiple origins of B chromosomes in fish genus *Characidium*. *PLoS ONE*, *9*, e107169.

Patawang, I., Phimphan S., Tanomtong, A. & Nithikulworawong, N. (2014). Rice frog, *Fejervarya limnocharis* (Anura, Ranidae) from Northeast Thailand. *Cytologia*, *79*, 141-150.

Paul-Prasanth, B., Matsuda, M., Lau, E. L., Suzuki, A., Sakai, F., Kobayashi, T. & Nagahama, Y. (2006). Knock-down of DMY initiates female pathway in the genetic male medaka, *Oryzias latipes*. *Biochem. Biophys. Res. Commun.*, *351*, 815-819.

Pendas, A. M., Moran, P., Freije, J. P. & Garcia-Vazquez, E. (1994). Chromosomal mapping and nucleotide sequence of two tandem repeats of Atlantic salmon 5S rDNA. *Cytogenet. Cell Genet.*, *67*, 31–36.

Perrin, N. (2009). Sex reversal: a fountain of youth for sex chromosomes? *Evolution*, *63*, 3043-3049.

Pita, S., Panzera, F., Sánchez, A., Palomeque, T. & Lorite, P. (2017). Chromosome painting in Triatomine insects reveals shared sequences between X chromosomes and autosomes. *J. Med. Entomol.*, *54*, 44-49.

Plohl, M., Luchetti, A., Mestrovic, N. & Mantovani, B. (2008). Satellite DNAs between selfishness and functionality: structure, genomics and evolution of tandem repeats in centromeric (hetero)chromatin. *Gene*, *409*, 72-82.

Plohl, M., Meštrović, N. & Mravinac, B. (2012). Satellite DNA evolution. *Genome Dyn.*, *7*, 126-152.

Ptacek, M. B., Gerhardt, H. C. & Sage, R. D. (1994). Speciation by polyploidy in tree frogs: multiple origins of the tetraploid, *Hyla versicolor*. *Evolution*, *48*, 898-908.

Pyron, R. A. & Wiens, J. J. (2011). A large-scale phylogeny of Amphibia including over 2800 species, and a revised classification of extant frogs, salamanders, and caecilians. *Mol. Phylogenet. Evol.*, *61*, 543-583.

Reeves, A. & Tear, J. (2000). *Micromeasure.* Version 3.3. Available from: URL: http://www.colostate.edu/Depts/Biology/ MicroMeasure.

Richard, G. F., Kerrest, A. & Dujon, B. (2008). Comparative genomics and molecular dynamics of DNA repeats in Eukaryotes. *Microbiol. Mol. Biol. Rev.*, *72*, 686-727.

Rodrigues, D. S., Rivera, M. & Lourenço, L. B. (2012). Molecular organization and chromosomal localization of 5S rDNA in Amazonian *Engystomops* (Anura, Leiuperidae). *BMC Genet.*, *13*, 17.

Rodrigues, P. C., Aguiar, Jr. O., Serpieri, F., Lima, A. P., Uetanebaro, M. & Recco-Pimentel, S. M. (2011). Chromosome analysis of five Brazilian species of poison frogs (Anura: Dendrobatidae). *J. Genetics*, *90*, 31-37.

Romanenko, S. A., Serdyukova, N. A., Perelman, P. L., Pavlova, S. V., Bulatova, N. S., Golenishchev, F. N., Stanyon, R. & Graphodatsky, A. S. (2017). Intrachromosomal rearrangements in rodents from the perspective of comparative region-specific painting. *Genes*, *8*, 215.

Ron, S. R., Santos, J. C. & Cannatella, D. C. (2006). Phylogeny of the túngara frog genus *Engystomops* (=*Physalaemus pustulosus* species group; Anura: Leptodactylidae). *Mol. Phylogenet. Evol.*, *39*, 392-402.

Ron, S. R., Toral, E., Rivera, M. & Terán-Valdez, A. (2010). A new species of *Engystomops* (Anura: Leiuperidae) from southwestern Ecuador. *Zootaxa*, *2606*, 25-49.

Rosa, C., Aguiar, O., Giaretta, A. A. & Recco-Pimentel, S. M. (2003). Karyotypic variation in *Megaelosia* (Anura, Hylodinae) with the first description of a B-chromosome in a leptodactylid frog. *Copeia*, 2003, 166-174.

Rosset, S. D., Baldo, D., Lanzone, C. & Basso, N. G. (2006). Review of the geographic distribution of diploid and tetraploid populations of the *Odontophrynus americanus* species complex (Anura: Leptodactylidae). *J. Herpetol.*, *40*, 465-477.

Rubtsov, N. B., Karamysheva, T. V., Andreenkova, O. V., Bochkaerev, M. N., Kartavtsevam, I. V., Roslik, G. V. & Borissov, Y. M. (2004). Comparative analysis of micron and macro B chromosomes in the Korean field mouse *Apodemus peninsulae* (Rodentia, Murinae) performed by chromosome microdissection and FISH. *Cytogenet. Genome Res.*, *106*, 289-294.

Ruiz-Ruano, F. J., Cabrero, J., López-Léon, M. D. & Camacho, J. P. M. (2016). Satellite DNA content illuminates the ancestry of supernumerary (B) chromosome. *Chromosoma*, *126*, 487-500.

Saba, N. & Tripathi, N. K. (2014). Preliminary cytogenetic study and report of ZZ/ZW sex chromosomes in the bullfrog, *Hoplobatrachus tigerinus* (Anura, Amphibia) from high altitude area of Jammu and Kashmir, India. *The Nucleus*, *57*, 55-59.

Saito, Y., Edpalina, R. R. & Abe, S. (2007). Isolation and characterization of salmonid telomeric and centromeric satellite DNA sequences. *Genetica*, *131*, 157-166.

Sangpakdee, W., Pimphan, S., Tengjaroenkul, B., Pinthong, K., Neeratanaphan, L. & Tanomtong, A. (2017). Cytogenetic study of three microhylid species (Anura, Microhylidae) from Thailand. *Cytologia*, *82*, 67-74.

Schartl, M., Schmid, M. & Nanda, I. (2016). Dynamics of vertebrate sex chromosome evolution: from equal size to giants and dwarfs. *Chromosoma*, *125*, 553-571.

Schempp, W. & Schmid, M. (1981) Chromosome banding in Amphibia. VI. BrdU-replication patterns in Anura and demonstration of XX/XY sex chromosomes in *Rana esculenta*. *Chromosoma*, *83*, 697-710.

Schmid, M. (1978). Chromosome banding in Amphibia. II. Constitutive heterochromatin and nucleolus organizer regions in Ranidae, Microhylidae and Rhacophoridae. *Chromosoma*, *68*, 131-148.

Schmid, M., Vitelli, L. & Batistoni, R. (1987). Chromosome banding in Amphibia. XI. Constitutive heterochromatin, nucleolus organizers, 18S+28S and 5S ribosomal RNA genes in Ascaphidae, Pipidae, Discoglossidae and Pelobatidae. *Chromosoma*, *95*, p. 271-284.

Schmid, M., Steinlein, C., Friedl, R., de Almeida, C. G., Haaf, T., Hillis, D. M. & Duellman, W. E. (1990). Chromosome banding in Amphibia. XV. Two types of Y chromosomes and heterochromatin hypervariabilty in *Gastrotheca pseustes* (Anura, Hylidae). *Chromosoma*, *99*, 413-423.

Schmid, M., Steinlein, C. & Feichtinger, W. (1992). Chromosome banding in Amphibia. XVII. First demonstration of multiple sex chromosomes in amphibians: *Eleutherodactylus maussi* (Anura, Leptodactylidae). *Chromosoma*, *101*, 284-292.

Schmid, M., Feichtinger, W., Steilein, C., Haaf, T., Schartl, M., Visbal-García, R., Manzanilla-Pupo, J. & Fernández-Badillo, A. (2002a). Chromosome banding in Amphibia. XXVI. Coexistence of homomorphic XY sex chromosomes and a derived Y-autosome translocation in *Eleutherodactylus maussi* (Anura, Leptodactylidae). *Cytogenet. Genome Res.*, *99*, 330-340.

Schmid, M., Feichtinger, W., Steinlein, C., Nanda, I., Mais, C., Haaf, T., Vísbal-Garcia, R. & Fernández-Badillo, A. (2002b). Chromosome banding in Amphibia. XXII. Atypical Y chromosomes in *Gastrotheca walkeri* and *Gastrotheca ovifera* (Anura, Hylidae). *Cytogenet. Genome Res.*, *96*, 228-238.

Schmid, M., Feichtinger, W., Steilein, C., Visbal-García, R. & Fernández-Badillo, A. (2003). Chromosome banding in Amphibia. XXVIII. Homomorphic XY sex chromosomes and a derived Y-autosome translocation in *Eleutherodactylus riveroi* (Anura, Leptodactylidae). *Cytogenet. Genome Res.*, *101*, 62-73.

Schmid, M. & Steinlein, C. (2003). Chromosome banding in Amphibia. XXIX. The primitive XY/XX sex chromosomes of *Hyla femoralis* (Anura, Hylidae). *Cytogenet. Genome Res.*, *101*, 74-79.

Schmid, M., Steinlein, C., Bogart, J. P., Feichtinger, W., León, P., La Marca, E., Díaz, L. M., Sanz, A., Chen, S. H. & Hedges, S. B. (2010). *The Chromosomes of Terraranan Frogs: insights into vertebrate cytogenetics*. Basel, SW. S. Karger AG, 559 p.

Schmid, M., Steinlein, C., Bogart, J. P., Feichtinger, W., Haaf, T., Nanda, I., del Pino, E. M., Duellman, W. E. & Hedges, S. B. (2012). The hemiphractid frogs. Phylogeny, embryology, life history, and cytogenetics. *Cytogenet. Genome Res.*, *138*, 69-384.

Schmid, M. & Steinlein, C. (2015). Chromosome banding in Amphibia. 32. The genus *Xenopus* (Anura, Pipidae). *Cytogenet. Genome Res.*, *145*, 201-217.

Session, A. M., Yoshinobu, U., Taejoon, K., Chapman, J. K., Toyoda, A., Takahashi, S., Fukui, A., Hikosaka, A., Suzuki, A., Kondo, M., Heeringen, S. J., Quigley, I., Heinz, S., Ogino, H., Ochi, H., Hellsten, U., Lyons, J. B., Simakov, O., Putnam, N., Stites, J., Kuroki, Y., Tanaka, T., Michiue, T., Watanabe, M., Bogdanovic, O., Lister, R., Georgiou, G., Paranjpe, S. S., Kruijsbergen, I, Shu, S., Carlson, J., Kinoshita, T., Ohta, Y., Mawaribuchi, S., Jenkins, J., Grimwood, J., Schmutz, J., MitroS, T., Mozaffari, S., Suzuki, Y., Haramoto, Y., Yamamoto, T. S., Takagi, C., Heald, R., Miller, K., Haudenschild, C., Kitzman, J., Nakayama, T., Izutsu, Y., Robert, J., Fortriede, J., Burns, K., Lotay, V., Karimi, K., Yasuoka, Y., Dichmann, D. S., Flajnik, M. F., Houston, D. W., Shendure, J., DuPasquier, L., Vize, P. D., Zorn, A.

M., Ito, I., Marcotte, E., Wallingford, J. B., Ito, Y, Asashima, M., Ueno, N., Matsuda, Y., Veenstra, G. J. V., Fujiyama, A., Harland, R. M., Taira, M. & Rokhsar, D. S. (2016). Genome evolution in the alloteraploid frog *Xenopus laevis*. *Nature*, *538*, 336-343.

Sharbel, T. F., Green, D. M. & Houben, A. (1998). B chromosome origin in the endemic New Zealand frog *Leiopelma hochstetteri* through sex chromosome evolution. *Genome*, *41*, 14-22.

Shibata, F., Hizume, M. & Kuroki, Y. (1999). Chromosome painting of Y chromosomes and isolation of a Y chromosome-specific repetitive sequence in the dioecious plant *Rumex acetosa*. *Chromosoma*, *108*, 266-270.

Silva, D. M. Z., Daniel, S. N., Camacho, J. P. M., Utsunomia, R., Ruiz-Ruano, F. J., Penitente, M., Pansonato-Alves, J. C., Hashimoto, D. T., Oliveira, C., Porto-Foresti, F. & Foresti, F. (2016). Origin of B chromosomes in the genus *Astyanax* (Characiformes, Characidae) and the limits of chromosome painting. *Mol. Genet. Genomics*, *291*, 1407-1418.

Sinclair, A. H., Berta, P., Palmer, M. S., Hawkins, J. R., Griffiths, B. L., Smith, M. J., Foster, J. W., Frischauf, A. M., Lovell-Badge, R. & Goodfellow, P. N. (1990). A gene from the human sex determining region encodes a protein with homology to a conserved DNA-binding motif. *Nature*, *346*, 240-244.

Singh, L., Purdom, F. & Jones, K. W. (1976). Satellite DNA and evolution of sex chromosomes. *Chromosoma*, *59*, 43-62.

Singh, L., Purdom, I. F. & Jones, K. W. (1980). Sex chromosome associated satellite DNA: evolution and conservation. *Chromosoma*, *79*, 137-157.

Slamovits, H. C. & Rossi, M. S. (2002). Satellite DNA: agent of chromosomal evolution in mammals. *Mastozoologia Neotropical*, *9*, 297-308.

Smith, C. A., Roeszler, K. N., Ohnesorg, T., Cummins, D. M., Farlie, P. G., Doran, T. J. & Sinclair, A. H. (2009). The avian Z-linked gene DMRT1 is required for male sex determination in the chicken. *Nature*, *461*, 267-271.

Steinemann, S. & Steinemann, M. (2005). Retroelements: tools for sex chromosome evolution. *Cytogenet. Genome Res.*, *110*, 134-143.

Stöck, M., Horn, A., Grossen, C., Lindtke, D., Sermier, R., Betto-Colliard, C., Dufresnes, C., Bonjour, E., Dumas, Z., Luquet, E., Maddalena, T., Sousa, H. C., Martinez-Solano, I. & Perrin, N. (2011). Ever-young sex chromosomes in European tree frogs. *PLoS Biol.*, *9*, e1001062.

Stöck, M., Savary, R., Zaborowska, A., Gorecki, G., Brelsford, A., Rozenblut-Kocisty, B., Ogielska, M. & Perrin, N. (2013a). Maintenance of ancestral sex chromosomes in Palearctic tree frogs: direct evidence from *Hyla orientalis*. *Sex. Dev.*, *7*, 261-266.

Stöck, M., Savary, R., Betto-Colliard, C., Biollay, S., Jourdan-Pineau, H. & Perrin, N. (2013b). Low rates of X-Y recombination, not turnovers, account for homomorphic sex chromosomes in several diploid species of Palearctic green toads (*Bufo viridis* subgroup). *J. Evol. Biol.*, *26*, 674-682.

Suárez, P., Cardozo, D., Baldo, D., Pereyra, M. O., Faivovich, J., Orrico, V. G. D., Catroli, G. F., Grabiele, M., Bernarde, P. S., Nagamachi, C. Y., Haddad, C. B. & Pieczarka, J. C. (2013). Chromosome evolution in Dendropsophini (Amphibia, Anura, Hylinae). *Cytogenet. Genome Res.*, *141*, 295-308.

Sun, Y. B., Xiong, Z. J., Xiang, X. Y., Liu, S. P., Zhou, W. W., Tu, X. L., Zhong, L., Wang, L., Wu, D. D., Zhang, B. L., Zhu, C. L., Yang, M. M., Chen, H. M., Li, F., Zhou, F., Feng, S. H., Huang, C., Zhang, G. J., Irwin, D., Hillis, D. M., Murphy, R. W., Yang, H. M., Che, J., Wang, J. & Zhang, Y. P. (2015). Whole-genome sequence of the Tibetan frog *Nanorana parkeri* and the comparative evolution of tetrapod genomes. *P. Natl. Acad. Sci. USA*, *112*, E1257-E1262.

Targueta, C. P., Rivera, M., Souza, M. B., Recco-Pimentel, S. M. & Lourenço, L. B. (2010). Cytogenetic contributions for the study of the Amazonian *Engystomops* (Anura; Leiuperidae) assessed in the light of phylogenetic relationships. *Mol. Phylogenet. Evol.*, *54*, 709-725.

Targueta, C. P., Rivera, M. & Lourenço, L. B. (2012). Karyotypic differentiation via 2n reduction and a finding of a case of triploidy in anurans of the genus *Engystomops* (Anura, Leiuperidae). *Genetica, 139*, 1339-1347.

Teixeira, L., Targueta, C. P., Seger, K. R., Orrico, V. D. G. & Lourenço, L. B. (2016). Comparative cytogenetics of tree frogs of the *Dendropsophus marmoratus* (Laurenti, 1768) group: conserved karyotypes and interstitial telomeric sequences. *Comp. Cytogenet., 10*, 753-767.

Terurel, M., Cabrero, J., Montiel, E. E., Acosta, M. J., Sánchez, A. & Camacho, J. P. M. (2009a). Microdissection and chromosome painting of X and B chromosomes in *Locusta migratoria. Chrom. Res., 17*, 11-18.

Teruel, M., Cabrero, J., Perfectti, F., Acosta, M. J. & Sánchez, A. (2009b). Microdissection and Chromosome Painting of X and B chromosomes in the Grasshoppers *Eyprepocnemis plorans*. *Cytogenet. Genome Res., 125*, 286-291.

Toledo, F. L., Siqueira, S., Duarte, T. C., Veiga-Menoncello, A. C. P., Recco-Pimentel, S. M. & Haddad, C. F. B. (2010). Description of a new species of *Pseudopaludicola* Miranda-Ribeiro, 1926 from the state of São Paulo, Southeastern Brazil (Anura, Leiuperidae). *Zootaxa, 2496*, 38-48.

Tomatis, C. G., Baldo, D., Kolenc, F. & Borteiro, C. (2009). Chromosomal variation in the species of the *Physalaemus henselii* group (Anura, Leiuperidae). *J. Herpetol., 43*, 555-560.

Ugarcović, D. & Plohl, M. (2002). Variation in satellite DNA profiles – causes and effects. *EMBO J., 21*, p. 5955-5959.

Uno, Y., Nishida, C., Oshima, Y., Yokoyama, S., Miura, I., Matsuda, Y. & Nakamura, M. (2008). Comparative chromosome mapping of sex-linked genes and identification of sex chromosomal rearrangements in the Japanese wrinkled frog (*Rana rugosa*, Ranidae) with ZW and XY sex chromosome systems. *Chromosome Res., 16*, 637–647.

Uno, Y., Nishida, C., Takagi, C., Ueno, N. & Matsuda, Y. (2013). Homoeologous chromosomes of *Xenopus laevis* are highly conserved after whole-genome duplication. *Heredity*, *111*, 430–436.

Uno, Y., Nishida, C., Takagi, C., Igawa, T., Ueno, N., Sumida, M. & Matsuda, Y. (2015). Extraordinary diversity in the origins of sex chromosomes in anurans inferred from comparative gene mapping. *Cytogenet. Genome Res.*, *145*, 218−229.

Vacher, J. P., Kok, P. J. R., Rodrigues, M. T., Lima, J. D., Lorenzini, A., Martinez Q., Fallet, M., Courtois, E. A., Blanc, M., Gaucher, P., Dewynter, M., Jairam, R., Ouboter, P., Thébaud, C. & Fouquet, A. (2017). Cryptic diversity in Amazonian frogs: Integrative taxonomy of the genus *Anomaloglossus* (Amphibia: Anura: Aromobatidae) reveals a unique case of diversification within the Guiana Shield. *Mol. Phylogenet. Evol.*, *112*, 158-173.

Veiga-Menoncello, A. C. P., Lourenço, L. B., Strussmann, C., Rossa-Feres, D. C., Andrade, G. V., Giaretta, A. & Recco-Pimentel, S. M. (2014). A phylogenetic analysis of *Pseudopaludicola* (Anura) providing evidence of progressive chromosome reduction. *Zool. Scripta*, *43*, 261-272.

Vences, M., Thomas, M., Van Der Meijden, A., Chiari, Y. & Vieites, D. R. (2005a). Comparative performance of the 16S rRNA gene in DNA barcoding of amphibians. *Front. Zool.*, *2*, 1-12.

Vences, M., Thomas, M., Bonett, R. M. & Vieites, D. R. (2005b). Deciphering amphibian diversity through DNA barcoding: chances and challenges. *Phil. Trans. R. Soc. B.*, *360*, 1859-1868.

Vicari, M. R., Pistune, H. F. M., Castro, J. P., Almeida, M. C., Bertollo, L. A., Moreira-Filho, O., Camacho, J. P. & Artoni, R. F. (2011). New insights on the origin of B chromosomes in *Astyanax scabripinnis* obtained by chromosome paining and FISH. *Genetica*, *139*, 1073-1081.

Vittorazzi, S. E., Lourenço, L. B., Del-Grande, M. L. & Recco-Pimentel, S. M. (2011). Satellite DNA derived from 5S rDNA in *Physalaemus cuvieri* (Anura, Leiuperidae). *Cytogenet. Genome Res.*, *134*, 101-107.

Vittorazzi, S. E., Lourenço, L. B. & Recco-Pimentel, S. M. (2014a). Long-time evolution and highly dynamic satellite DNA in leptodactylid and hylodid frogs. *BMC Genet.*, *15*, 111.

Vittorazzi, S. E., Quinderé, Y. R. S. D., Recco-Pimentel, S. M., Tomatis, C., Baldo, D., Lima, J. R. F., Ferro, J. M., Lima, J. D. & Lourenço, L. B. (2014b). Comparative cytogenetics of *Physalaemus albifrons* and *Physalaemus cuvieri* species groups (Anura, Leptodactylidae). *Comp. Cytogenet.*, *8*, 103-123.

Vittorazzi, E., Lourenço, L. B. & Recco-Pimentel, S. M. (2016). Chromosomal analysis of *Physalaemus kroyeri* and *Physalaemus cicada* (Anura, Leptodactylidae). *Com. Cytogenet.*, *10*, 311-323.

Volff, J. N., Nanda, I., Schmid, M. & Schartl, M. (2007). Governing sex determination in fish: regulatory putches and ephemeral dictators. *Sex. Dev.*, *1*, 85-99.

Wang, X., Zhang, Q., Ren, J., Jiang, Z., Wang, C., Zhuang, W. & Zhai, T. (2009). The preparation of sex-chromosome-specific painting probes and construction of sex chromosome DNA library in half-smooth tongue sole (*Cynoglossus semilaevis*). *Aquaculture*, *297*, 78-84.

Warburton, P. E., Dolled, M., Mahmood, R., Alonso, A., Li, S., Naritomi, K., Tohma, T., Nagai, T., Hasegawa, T., Ohashi, H., Govaerts, L. C., Eussen, B. H., Van Hemel, J. O., Lozzio, C., Schwartz, S., Dowhanick-Morrissette, J. J., Spinner, N. B., Rivera, H., Crolla, J. A., Yu, C. & Warburton, D. (2000). Molecular cytogenetic analysis of eight inversion duplications of human chromosome 13q that each contain a neocentromere. *Am. J. Hum. Genet.*, *66*, 1794-1806.

Wu, G. F. & Zhao, E. M. (1985). Preliminary studies on karyotypes of the genus *Amolops* of the Hengduan Mountains. *Acta Herpetol. Sinica*, *4*, 276-282.

Yamada, K., Nishida-Umehara, C. & Matsuda, Y. (2004). A new Family of satellite DNA sequences as a major component of centromeric heterochromatin in owls (Strigiformes). *Chromosoma*, *112*, 277-287.

Yoshimura, A., Nakata, A., Mito, T. & Noji, S. (2006). The characteristics of karyotypes and telomeric satellite DNA sequences in the cricket, *Gryllus bimaculatus* (Orthoptera, Gryllidae). *Cytogenet. Genome Res.*, *112*, 329-336.

Yoshimoto, S., Okada, E., Umemoto. H., Tamura, K., Uno, Y., Nishida-Umehara, C., Matsuda, Y., Takamatsu, N., Shiba, T. & Ito, M. (2008). A W-linked DM-domain gene, DM-W, participates in primary ovary development in *Xenopus laevis*. *P. Natl. Acad. Sci. USA*, *105*, 2469-2474.

In: An Essential Guide to Cytogenetics ISBN: 978-1-53613-370-7
Editors: N. Norris and C. Miller 

*Chapter 2*

# HUMAN CYTOGENETIC BIOMONITORING

***Armanda Teixeira-Gomes[1,2], Solange Costa[1,2,*]***
***and João Paulo Teixeira[1,2]***
[1]EPIUnit – Instituto de Saúde Pública,
Universidade do Porto, Porto, Portugal
[2]Environmental Health Department,
National Institute of Health, Porto, Portugal

## ABSTRACT

Humans are daily exposed to a variety of potentially harmful agents in the air they breathe, liquids they drink, food they eat and products they use. Long-standing evidence of the bond between health and environment has led to the recognition for the need of sustainable development. On the other hand, there is an increasing global awareness of the inevitable limits of individual health care and the need to complement such services with effective public health strategies. According to World Health Organization (WHO), cancer is a leading cause of death worldwide. Additionally, at least 200 000 people die every year from occupational or work-related cancers. Exposure assessment aims at prevention. Establishing the health effects of various activities and exposures requires information about the levels of exposure and the biological effects

[*] Corresponding Author Email: solange.costa2@gmail.com.

resulting from the interaction between the organism and the chemical agent. The resulting data provides a basis for designing effective prevention and mitigation strategies. Cytogenetic endpoints have long been applied in surveillance of human genotoxic exposure and early effects of genotoxic carcinogens. Assays measuring chromosomal aberrations (CAs), micronucleus (MN) and sister chromatid exchange (SCE) in lymphocytes are well-established techniques extensively used in human biomonitoring studies to assess DNA damage at the chromosomal level. The relevance of cytogenetic alterations as a cancer risk biomarker is further supported by epidemiologic data linking CAs and MN with cancer risk in human populations. Thus, the use of cytogenetic markers in human biomonitoring is of paramount importance due to its predictability regarding deleterious effects resulting from the exposure to environmental stressors.

**Keywords:** human biomonitoring, cytogenetic, chromosomal aberrations, micronucleus and sister chromatid exchange

## AN INTRODUCTION TO HUMAN BIOMONITORING

Humans are exposed to several potential harmful agents in their daily life activities, in the air, water, soil, food and products they use [1]. Identification, evaluation and control of environmental hazards is of extreme importance to prevent the health effects of exposure to substances. Human biomonitoring is frequently used to provide early warning signals of excessive exposure, allowing the measure of the substances themselves, their metabolites or markers of subsequent health effects in body fluids or tissues [2]. Studying adverse health outcomes related to the environmental exposures (in the living and working environment) is a major societal challenge today.

Exposure can be assessed by biological monitoring, with measures of biological endpoints (biomarkers or biological indicators). Biological monitoring assesses the interaction of potential harmful agents with the biological systems by analyzing chemicals, metabolites, enzymes and other biochemical substances in tissues, body fluids or other accessible samples [1]. Other method used to protect human health in case of exposure to potential harmful substances is called environmental monitoring, that encompasses the measure of those substances in environmental matrices like air, water, soil, food, etc. Environmental monitoring is particularly important to the determination of the sources of exposure. The two methods, biological and environmental monitoring, complete each other and are commonly used simultaneously, since they provide different but complementary information [3, 4]. Nevertheless, human biomonitoring is essential because it allows a better estimation of the internal dose of a compound (dose really taken up) and consequently, its potential health risks, despite the route of exposure - inhalation, absorption through the skin, and ingestion.

Human biomonitoring depends on the use of biomarkers that are defined as a xenobiotically induced alteration in cellular or biochemical components or processes, structures, or functions measurable in a biological system or sample [5]. A good biomarker must be sensitive, specific, relevant, reproducible and measurable in the population. Biomarkers are traditionally divided into three main categories - biomarkers of exposure, susceptibility and effect. Biomarkers of exposure are defined as the compound, parent compound or its metabolites, or the product of an interaction between the xenobiotic and its target molecule or cell that is measured in a compartment within an organism (e.g., lead levels in blood; arsenic levels in hair). Biomarkers of susceptibility are indicators of an inherent or acquired capacity of an organism to respond to the challenge of exposure to a specific chemical (e.g., polymorphic genes of metabolizing enzymes). Biomarkers of effect are measurable biochemical, physiological, behavioral, functional, or other alterations within an organism induced by the exposure and depending on the magnitude of the alterations can be recognized as associated with an established or possible health impairment or disease [1, 6].

Biomarkers of exposure are specific for the identification of the agent of exposure, while the biomarkers of effect are usually unspecific for the causative agent [6]. Although they are less specific in the identification of the agent of exposure, they can be more predictive of the induced toxicity. They are biological indicators of the response of the body to exposures indicating early sub-clinical changes, which can develop into pathological consequences [7]. Biomarkers of effect either indicate early processes that precede a disease or predict the development and presence of disease [8]. The cytogenetic biomarkers that are going to be addressed further in this chapter belong to the category of biomarkers of effect, and are extensively used tools in human biomonitoring studies, particularly in occupational studies, to evaluate the genotoxic potential of the exposure to hazardous chemicals [9].

## Cytogenetic Biomarkers in Human Biomonitoring

Cytogenetic biomarkers are extensively used tools in human population studies to assess the impact of environmental, occupational and medical factors on genomic stability [9]. Nowadays, the most used cytogenetic endpoints in human biomonitoring studies are chromosomal aberrations (CAs) and micronucleus assay (MN). The increased frequency of these indicators in exposed workers is a consistent finding in several studies [10-17]. Also, the interest in assays measuring these cytogenetic alterations has increased since they have been linked with cancer risk in human epidemiologic studies [18, 19].

One critical issue in the use of cytogenetic endpoints in human biomonitoring is the selection of the biological specimen to investigate [20]. These endpoints are usually assessed in readily available surrogate cells to estimate events occurring in target tissues and to provide early warning signals for adverse health effects. Despite the documented usefulness of non-blood cells (e.g., buccal, nasal, urothelial) in biomonitoring [21] the peripheral blood lymphocytes (PBLs) are the most frequently used cells in human studies. The major reason to use lymphocytes is because they can be damaged in any tissue/organ specific toxic environment, since these cells circulate throughout the body and have reasonably long life-span [22]. The preference in the use of lymphocytes is also given to the fact that they exist in large number in the human body and are relatively simple to harvest.

In this chapter we are going to review the mechanisms and methodology of the major cytogenetic endpoints used in human biomonitoring: CAs, MN and sister chromatid exchanges (SCEs). An overview of these cytogenetic assays can be observed in Figure 1.

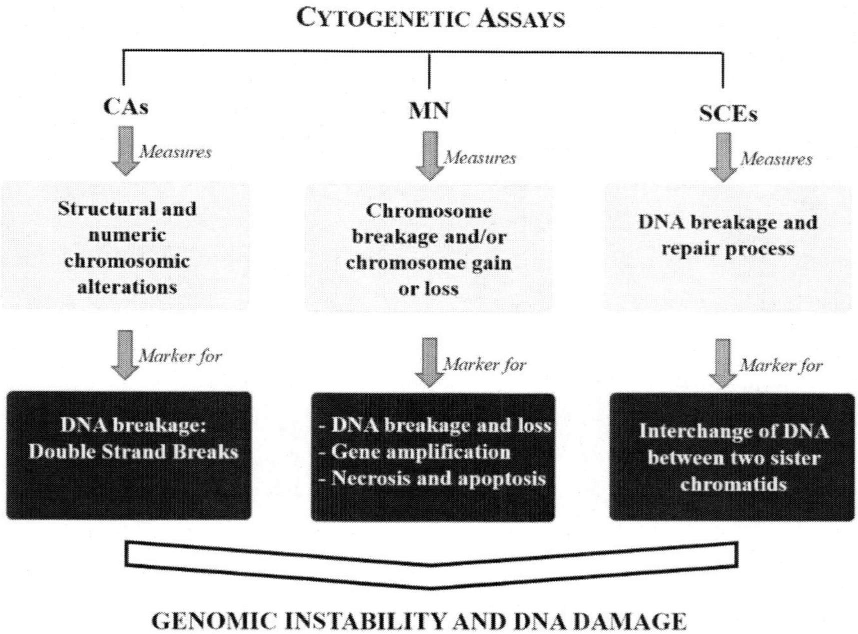


Figure 1. Overview of the cytogenetic assays.

## Chromosomal Aberrations (CAs): Mechanisms and Methodology

Chromosomal aberrations (CAs) are changes in normal chromosome structure or number that can occur as a consequence of exposure to chemicals or radiation [23]. Chromosome structural changes have been assessed for decades by CA assay in PBLs to evaluate the exposure to contaminants in occupational and environmental settings.

Structural CAs may be induced by direct DNA breakage, replication on a damaged DNA template, inhibition of DNA synthesis or other mechanisms (e.g., topoisomerase II inhibitors) [23]. The formation of structural CAs requires one or several double-strand breaks (DSBs). Mechanistic and morphological differences categorize this CAs in two types: chromosome-type aberrations and chromatid-type aberrations. Concerning the morphology, chromosome-type aberrations involve both sister chromatids of a chromosome or more chromosomes, and chromatid-type aberrations only one chromatid of one or several chromosomes [24]. Regarding the mechanism, chromosome-type aberrations and chromatid-type aberrations differ in the type of initial lesion (induced by different agents) and the DNA repair mechanisms that follow. Those lesions are commonly repaired by three main mechanisms: homologous recombination (HR), non-homologous end joining (NHEJ) repair and single-strand annealing (SSA). HR is an error-free process, highly accurate that precisely restores the original sequence at the break. NHEJ is an error-prone repair process that joins directly the two broken ends which usually leads to small-scale alterations at the break site. SSA mainly leads to the formation of interstitial deletions [25].

Chromosome-type aberrations are usually generated *in vivo* by S-phase independent clastogens in resting $G_0/G_1$ lymphocytes. After DNA synthesis and chromosome duplication, the aberrations formed are doubled and chromosome-type breaks and exchanges (e.g., dicentric and ring chromosomes) are seen in metaphase. The duplication originates symmetric lesions on both sister chromatids. They reflect DSBs that are incompletely repaired or unrepaired by NHEJ repair and SSA.

Chromatid-type aberrations arise predominantly *in vitro* during the S-phase of the cultured lymphocytes. These type of breaks and exchanges are the response to base modifications and single-strand breaks (SSBs) induced *in vivo* by S-phase dependent clastogens. DSBs are a consequence of these lesions during DNA replication. Incomplete or failed repair of these lesions by conservative HR will generate chromatid-type aberrations formation in a subsequent metaphase [26, 27]. Chromatid-type aberrations formation is the mechanism most likely to be applicable to human biomonitoring studies since it represents the common process by which clastogenic chemicals induce CA.

As previously mentioned, the detection of structural CAs in human biomonitoring studies is performed in lymphocytes using the CA assay. This assay is the most widely used and best validated because the mechanisms for CAs induction are better understood and most environmental toxic substances have been shown to induce structural CAs [22, 28]. In fact, the CA assay is recognized as one of key tests for genotoxic compounds, being its protocol defined by OECD guidelines for the testing of chemicals (OECD TG 473) [29].

Lymphocytes are cultured in the presence of a mitogen, for instance phytohemagglutinin (PHA), in order to stimulate cell division. A spindle inhibitor (e.g., colcemid, colchicine) is added to the culture after a determined period to stop the replication when cells are in the metaphase. A schematic of the assay is represented in Figure 2.

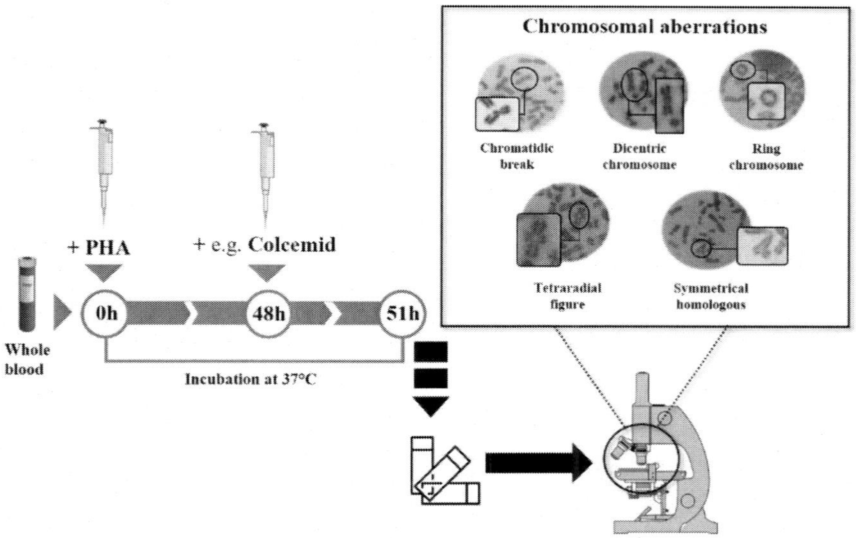


Figure 2. Schematic representation of CAs assay described by Roma-Torres et al., 2006 [31] and examples of microscopical observations.

Afterwards, the metaphase-arrested cells are ready for scoring. This method is not appropriate for the estimation of numerical CAs, since chromosome loss can occur. Numerical CAs are changes in normal chromosome number, a consequence of abnormal cell division. Cells are considered aneuploid when they contain extra (hyperploid) or missing (hypoploid) chromosome. These changes may occur due to damage to the mitotic spindle and associated elements, damage to chromosome sub-structures, alterations in cell physiology and mechanical disruption [26, 30]. The additional use of fluorescence in situ hybridization (FISH) methods that have higher sensitivity and efficiency, enables the detection of subtle aberrations, as chromosome-specific rearrangements and/or loss, which are not apparent when using the conventional Giemsa staining [24].

## Micronucleus (MN): Mechanisms and Methodology

Micronucleus (MN) are extra-nuclear bodies resultant from damaged chromosome fragments and/or whole chromosomes that are not included in the daughter nuclei at completion of telophase during mitosis. The chromosome fragments or whole chromosomes lag behind during the segregation process in anaphase because they did not attach properly to the spindle. The DNA material is eventually enclosed by a nuclear membrane being morphologically similar to nuclei (with the exception of its smaller size) after conventional nuclear staining [32].

Chromosomal fragments can result from direct DNA breakage, conversion of SSBs into DSBs during replication, or inhibition of DNA synthesis [26]. Whole chromosomes are generally formed from defects in the chromosome segregation machinery, such as deficiencies in the cell cycle checkpoint genes, dysfunction of the mitotic spindle, defects in kinetochore or other parts of the mitotic activity [32]. MN formation can also be related to damage to chromosomal sub-structures, hypomethylation of centrometric and pericentrometric DNA or mechanical disruption [26, 33]. Hence, MN may represent a measure of both chromosome breakage and chromosome loss, so an increased frequency of micronucleated cells can reflect exposures to clastogenic (e.g., strutural CAs) or aneugenic (e.g., numerical CAs) agents.

The MN assay can also provide parameters of genetic damage [e.g., nucleoplasmic bridge (NPB) and nuclear buds (NBUDs)] apart from the chromosome breakage and loss. DSBs misrepair may result in the development of a dicentric chromosome and an acentric fragment. Typically, at anaphase the dicentric chromosome forms a NPB between the two daughter nuclei, when its centromeres are pulled to the opposite poles of the cell, and the acentric fragment results in a micronuclei [32, 33]. NBUDs are characterized by having the same morphology as a micronuclei, but they are attached to the nucleus by a narrow or wide nucleoplasmic connection [33]. The nuclear budding takes place during S-phase of the cell cycle and cells are sometimes referred as "broken egg" cells. Nuclear budding seems to be the mechanism by which cells remove amplified and/or excess DNA, being NBUDs considered a biomarker of gene amplification and/or altered gene dosage [32].

Once formed, a micronucleated cell can be eliminated by apoptosis, the MN can be expelled from the cell, retained in the cytoplasm or incorporated into the main nucleus. However, the post-mitotic fate of the MN and the micronucleated cell is still poorly understood [20].

Currently, MN are extensively used as genotoxicity biomarker and has found a place in human biomonitoring studies, being one of the best validated techniques to evaluate chromosomal damage in humans [34]. In fact, MN assay offers more advantages than SCE and CA assays - MN scoring is simpler, requires shorter training and is less time consuming.

MN frequency can be assessed in exfoliated epithelial cells (buccal, nasal mucosa, urine), erythrocytes [OECD TG 474 [35]] or in lymphocytes. The *in vitro* mammalian cell MN test protocol for testing chemicals, encompassing cultured primary human or other mammalian PBLs, was adopted in 2010 by the OECD (OECD TG 487) [36]. In human population studies, the PBLs are the most commonly used tissue and the most frequently used test the cytokinesis-block micronucleus (CBMN) assay. This assay has been extensively validated worldwide for assessing environmental effects on chromosome damage in blood and epithelial tissues in human populations. The MN frequencies can be influenced by age, sex, nutritional status or smoking habits [37, 38]. A schematic representation of the CBMN assay is presented in Figure 3.

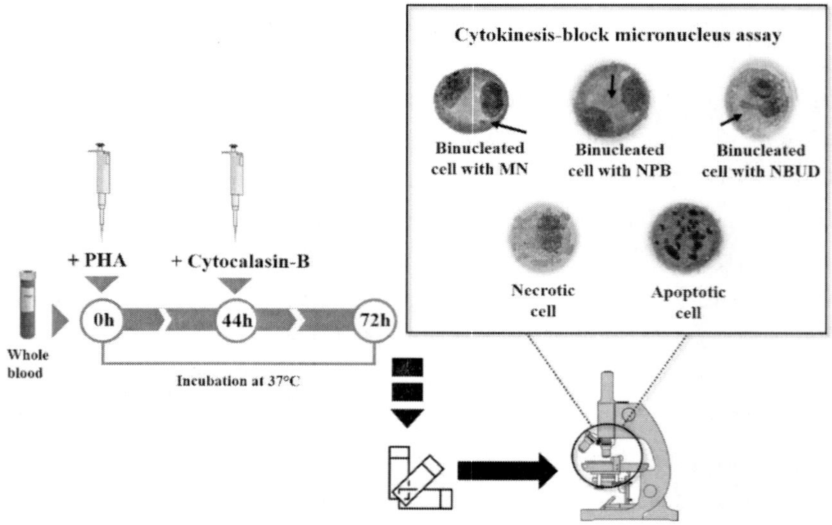


Figure 3. Schematic representation of CBMN assay described by Costa et al., 2006 [12] and examples of microscopical observations.

Like the CA assay, CBMN assay is performed in cultured PBLs that have divided once and were stimulated to proliferate in the presence of a mitogen (PHA). MN can only be expressed in cells that have completed one nuclear division, and for that reason is necessary to differentiate these cells from resting cells to assess the cytogenetic damage [23]. Cytochalasin-B (actin-polymerization inhibitor) is added before the first mitosis to block the cytokinesis allowing the identification of the once-divided cells in culture (binucleated cells) and the cells that did not divide or escaped the blocking of cytokinesis (mononucleated cells). Of note that only the cells that completed one nuclear division during *in vitro* culture can express the damage inflicted *in vivo*. The presence of MN in binucleated cells corresponds to pre-existing *in vivo* chromosome damage that have accumulated in the DNA since the last replication and lesions expressed as MN during the first *in vitro* mitosis. In mononucleated cells, MN are a result of genomic instability accumulated over many years in stem cells and circulating lymphocytes. The CBMN test also provides information on the nuclear division index by the registration of mono, bi, tri and tetranucleated cells, offering useful information regarding the cytostatic potential of exposures by the identification of compounds that stimulate or delay cell division [39].

The combination of the CBMN assay with FISH probes contributes to the higher specificity and sensitivity of the method, since it allows the differentiation of MN containing a whole chromosome (centromere positive MN), resulting from the action of aneugenic agents, or an acentric chromosome fragment (centromere negative MN), reflecting the exposure to clastogenic agents.

In conclusion, besides aneugenic and clastogenic events this assay can provide information on several genotoxic and cytotoxic markers, such as, NPB (marker of chromosome rearrangement), NBUDs (marker of gene amplification) and nuclear division index (estimation of cell division inhibition).

## Sister Chromatid Exchanges (SCEs): Mechanisms and Methodology

Sister chromatid exchanges (SCEs) are the result of interchanges of DNA between two sister chromatids in a duplicating chromosome. These reciprocal DNA exchanges take place during the S-phase of the cell cycle, when DNA is duplicated, resulting in two identical daughter chromatids. The effective breakage of DNA in sister chromatids followed by the reunion of the chromatids with one another, apparently at the same location, results in the SCEs [40]. Even though the mechanism of formation of SCE is not completely understood, they seem to be a consequence of the replication of a damaged DNA template, possibly at the replication fork [26]. The likely pathway for SCEs formation involves the initial collapse of a replication fork when facing a previously existent gap or SSB in one parental strand [41]. Subsequently, the formation of a replication-associated DSB with one free end takes place, initiating the repair process by conservative HR with the invasion of the intact sister strand to serve as a template for DNA synthesis. Resolution of the resulting Holliday junction by non-crossing over will follow, triggering the induction of SCEs [42] (Figure 4). Another possible mechanism for SCEs formation is the initial stall of the replication fork due to the existence of obstacles, namely adducts, on the DNA template. The cleavage of these intermediary structures by endonucleases will induce DSBs with one free end that can be further repaired by conservative HR as previously described, resulting in SCEs [42].

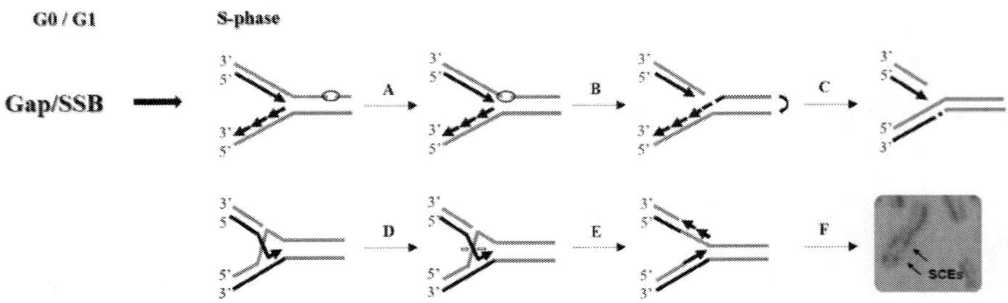


Figure 4. Mechanism of SCEs formation. A) Replication fork approaches a SSB. B) Collapsed replication fork; fork breaks; repair synthesis. C) 5' to 3' double-strand break resection. D) HR pathway, strand invasion. E) Resolution of the Holliday junction in the orientation shown by the arrows results in SCE, as illustrated by the grey/black color junctions in the new "parental" strands. F) The replication fork is restored (taken from Costa et al., 2013 [43]).

Taylor and coworkers were the first observing SCEs with the use of autoradiography on cells that underwent a cycle of triatiated thymidine incorporation followed by a replication cycle in nonradioactive medium [44]. Nowadays, 5-bromo-2′-deoxyuridine (BrdU), an analogue of DNA thymidine, is used for a better visualization of the SCEs. A standard fluorescence assay is used based on differential staining of the sister chromatids after inducing cell division in the presence of BrdU [45]. BrdU is efficiently incorporated into the DNA during replication due to the closer resembles with thymidine. This technique takes advantage of the semiconservative nature of the DNA replication and when cells are cultured through a single replication cycle with the presence of BrdU in the medium, one DNA strand in each daughter chromatid is substituted by this thymidine analogue. After the second replication round the two sister chromatids differ in BrdU presence: one chromatid contains one substituted DNA strand, while both strands of its sister chromatid are substituted. The chromatids can be further differentiated with Hoechst dye, which fluoresces at lower intensity when bound to DNA substituted with BrdU than when bound to unsubstituted DNA [46]. Photosensitization with UV light is conducted, resulting in the selective degradation of the highly substituted chromatid, since it seems to preferentially break the bonds present in non-histone proteins of the chromatid containing more BrdU. Giemsa staining follows and the asymmetric distribution of BrdU-substituted DNA in sister chromatids can be observed by light microscopy, because brdU incorporation leads to a much weaker staining being the chromatid with more BrdU lighter in appearance [41]. SCE frequencies can afterwards be scored using a semi-automated computer-based metaphase finder. In Figure 5 a schematic representation of SCEs assay can be observed.

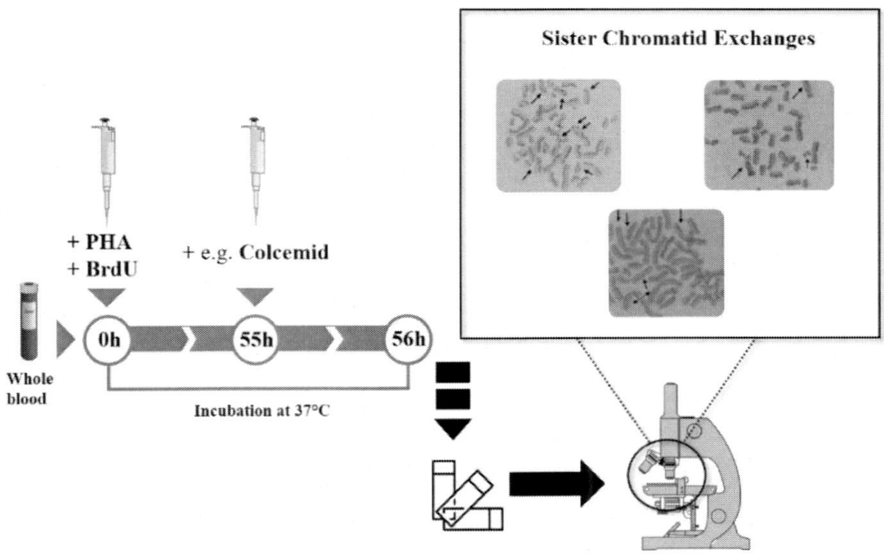


Figure 5. Schematic representation of SCEs assay described by Teixeira et al., 2004 [47] and examples of microscopical observations.

# CAs, MN, SCEs and Risk Predictivity of Human Diseases

Environmental exposures related to our lifestyle, infections and the occupation that we have throughout our life are considered a major cause of cancer in humans. According to the World Health Organization (WHO), cancer is a leading cause of death worldwide, being responsible for 8.8 million deaths in 2015. Globally, one in six deaths is due to cancer. It has been estimated that 3 to 6% of all cancers in the world are resultant from occupational exposures. Additionally, at least 200 000 people die every year from occupational or work-related cancers [48].

Exposure assessment aims at prevention. Looking at these numbers, it is of extreme importance to identify hazardous environmental substances and predict the development of cancer. Thus, a great challenge in biomarker research is the identification of long-term risks of developing cancer and the identification of early markers of occupational and environmental toxicity. The use of cytogenetic markers in human biomonitoring is of paramount importance due to its predictability regarding deleterious effects in health related with exposure to environmental stressors. Regarding carcinogenicity, the most frequently used biomarkers reflect genotoxic effects. Accordingly, these cytogenetic endpoints have been applied either in the surveillance of human exposure or in the investigation of the genotoxic potential of new compounds by the U.S. National Toxicology Program (NTP) and the International Agency for Research on Cancer (IARC). Epidemiologic data linking high frequencies of CAs and MN in PBLs with cancer risk in human populations supports the relevance of cytogenetic alterations as cancer risk biomarkers [18, 19].

SCEs were efficiently induced by several mutagenic or carcinogenic agents, particularly those that result in covalent DNA adducts or interfere directly or indirectly with DNA replication [45]. This assay has also gained popularity when it was observed that the cells that derived from patients with Bloom's syndrome, characterized by a high propensity for cancer, exhibit a hallmark high occurrence of SCE [49] due to dysregulation of HR by a defective helicase function [41]. Defects in HR process might be the cause of genetic instability, increased SCE and high incidence of cancer in early life associated with this disorder. Actually, increases in SCE frequency have been used as a biomarker in the clinical diagnose of Bloom's syndrome [41, 50].

SCEs is considered a good biomarker of effect to genotoxic/mutagenic agents, although no association between high SCEs frequencies and cancer risk was observed [18, 28, 51-53]. In the early 1990s, the preliminary reports from the Nordic Study Group on the Health Risk of Chromosome Damage mentioned the lack of association between the frequency of SCE and the incidence of cancer in humans [52, 54]. The results obtained in these initial studies were later confirmed in 1998 by a European cohort study [18] and in 2004 in a study from Bonassi and coworkers [53]. An association between SCEs level and cancer risk is difficult to evaluate because the baseline levels of SCEs are different among individuals and across studies [51]. Consequently, the use of SCE as biomarker for human biomonitoring has been declining, and CAs and MN are the preferred methods nowadays.

As previously mentioned, CAs have been associated with cancer risk prediction, being the most extensively used biomarker in human populations exposed to genotoxic agents [26]. The first reports regarding possible associations between the frequency of CAs and the incidence of cancer in humans emerged in the early 1990s, as for SCEs. Several studies with Italian and Nordic cohorts reported that the CAs levels in lymphocytes revealed to be predictive of overall cancer risk in humans [18, 52, 55, 56]. In fact, one of the latest studies mentioned, a report from 1998 of the European Study Group on Cytogenetic Biomarkers and Health, had the following results using both cohorts that comprised 3541 subjects examined between 1970 and 1988: subjects with higher levels of CAs presented an elevated standardized incidence ratio for cancer of 1.53 for the Nordic cohort and a mortality ratio of 2.01 for the Italian cohort demonstrating an increased cancer incidence in healthy individuals [18]. Another report for the referred European Group using both cohorts confirmed that the increased risk of cancer was related to high levels of CAs in lymphocytes, but independent of the exposure to carcinogenic agents [57]. Although, a subsequent study from Smerhovsky and coworkers that analyzed data from 3973 subjects reported a strong association between the frequency of CAs and the incidence of cancer in workers exposed to radon – an increase of 1% in the frequency of CAs was associated with an increase in cancer risk of 64% [58]. Nevertheless, the data collected in this study could not assess this association with exposure to other chemicals. Later, other study using the Nordic and Italian cohorts concluded that both chromosome-type aberrations and chromatid-type aberrations have the similar cancer risk predictive values [27], however some studies showed stronger association of chromosome-type aberrations with cancer risk predictivity [51]. A later study from 2007 of Boffetta and coworkers in a cohort with total of 6430 subjects from 9 laboratories in central Europe confirmed these previous results, observing relative risks of cancer of 1.78 and 1.81 for the medium and high tertiles levels of CAs [59]. Also, this study assessed the relation between the frequency of CAs and the risk of specific cancers and showed a relation between CAs and stomach cancer and possibly the colon and rectum cancers, regarding the lung and breast cancers the results were only suggestive of an association. A study from Bonassi and coworkers helped to clarify some of the referred issues, using a larger pooled analysis with 22 358 cancer-free subjects that underwent genetic screening and with follow-up time for cancer incidence of 10.1 years [60]. The following results were observed: an increased relative risk of cancer was found for high levels of CAs, this increase was mostly driven by chromosome-type aberrations, the strongest association was found for the stomach cancer and the effect of CA levels on overall cancer risk was not altered by exposure. These results were in general consistent with previous studies. A recent study in patients diagnosed with colorectal, lung and breast cancer also reported that CAs serve as predictive markers of cancer but in this particular study for the colorectal cancer association, only chromatid-type aberrations were significantly elevated [61]. The results obtained in all studies highlight the importance of CAs as biomarkers of cancer risk prediction in humans.

MN can be a result of chromatid-type aberrations occurring during the replication of DNA on a damaged template or chromosome-type aberrations initiated before mitosis and duplicated at replication [26]. Hence, taking into consideration the cancer risk predictivity of CAs and the mechanistic similarities between CAs and MN formation, it was expected that MN would also be possible biomarkers of cancer predictivity [20]. In the 1990s the first studies performed in this regard found no relation between MN frequency in PBLs and the cancer risk in humans [18, 52, 56]. Nevertheless, later in 2007 Bonassi and coworkers reported that MN formation is associated with early events in carcinogenesis [19]. The results emerged from a cohort of 6718 individuals that were screened for MN frequency between 1980 and 2002 and a significant increase of all cancers incidence was found in otherwise healthy subjects with an elevated frequency of MN in PBLs. The link between MN induction in PBLs and cancer development was further confirmed by case control and meta-analysis studies [62, 63]. A significantly higher MN frequency was observed in lymphocytes of individuals who developed cancer within 14 years after blood sampling (cases; $4.7 \pm 3.4$ MN/1000 BN cells) as compared to those who were still cancer free at the end of the follow-up period (controls; $1.5 \pm 1.7$; $p < 0.0001$) [62]. Additionally, subjects with a high MN frequency (>2.5 MN/1000 BN cells) had an increased risk of cancer death when compared with those with low frequency of MN (≤2.5 MN/1000 BN cells). A meta-analysis with findings from 37 publications about the association between MN frequency in PBLs and the presence of cancer diagnosis revealed a 28 to 64% increase in the baseline MN frequency of untreated cancer patients when compared to cancer-free referents [63]. Other recent studies using the CBMN assay reported an association between the increased MN frequencies in lymphocytes and the risk of colorectal and bladder cancers [64, 65]. A study from Maffei and coworkers found an increase of MN frequency in PBLs of patients with colorectal cancer ($16.82 \pm 6.56$) when compared to controls ($8.00 \pm 1.77$) [64]. This report also mentioned that the MN frequency found for the control group was in accordance with the MN frequency of healthy controls, adjusted for age and gender, of their laboratory database ($7.54 \pm 1.74$). Recently, an increased frequency of MN ($9.51 \pm 4.73$ versus $7.73 \pm 3.91$) and NBUDs ($5.06 \pm 3.36$ versus $3.99 \pm 2.44$) was observed in bladder cancer patients compared to controls [65]. In addition to the cancer predictivity of MN, other biomarkers of the CBMN assay (i.e., NPB and NBUDs) have been reported as stronger predictors of cancer risk [65-67]. All the recently accumulated data on the cancer predictive value of elevated MN frequencies makes the CBMN assay a good candidate for wide application in human biomonitoring studies.

In Figure 6 a chronological representation of the major findings regarding the aforementioned cytogenetic biomarkers and the cancer risk predicitivity in humans can be observed.

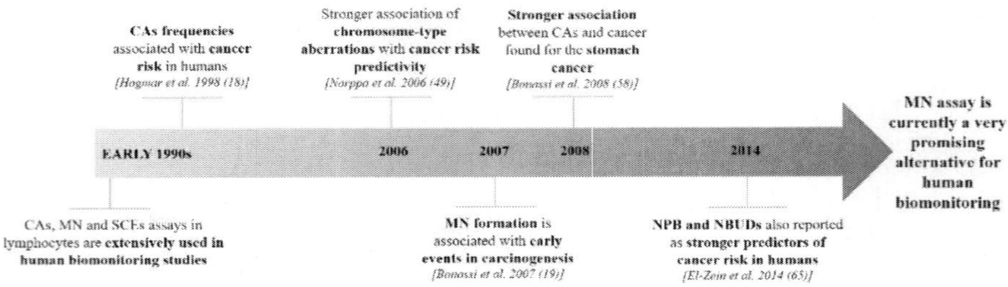


Figure 6. Chronological representation of the major findings regarding cytogenetic biomarkers and cancer risk predictivity.

## APPLICATION IN OTHER HUMAN CELLS

Other cell types widely used to assess the formation of MN in human biomonitoring studies are the exfoliated cells, such as buccal, nasal or urothelial cells. These non-blood cells have some advantages when compared with lymphocytes for the use in large human biomonitoring studies: are easier to collect by non-invasive procedures; and, in some cases, these tissues are the actual target of the carcinogenic agents of exposure. For instance, they are in immediate contact with genotoxic agents by inhalation and ingestion, also kidney and bladder cells are in contact with the metabolites of the agents of exposure [68].

An advantage for the use of these types of cells is the elimination of the culture step, since these are usually rapidly dividing tissues. Cells are collected (swabbed in the case of buccal and nasal cells and isolated from urine in the case of urothelial cells) and dropped or smeared into the slides and stained with the appropriate dye [69].

Increases in MN frequencies in exfoliated cells in response to specific exposures have been widely reported. Some examples are the increases in MN frequencies in buccal, nasal and urothelial cells of populations exposed to formaldehyde [70-74], arsenic [75-77] and smoking [78-82].

Apart from the use of these cells in occupational and environmental biomonitoring studies, they have also been related with risk predictivity of human diseases [83]. In 2000, Bloching and coworkers suggested that the use of MN assay in buccal mucosa could be important in the prediction of cancer risk in the upper aerodigestive tract [84]. Exfoliated buccal cells collected from breast and uterus of untreated cancer patients presented a significant increase in MN number when compared to control subjects [85]. A study performed in 45 cancer patients with lung, stomach or colorectal cancer found significant increases in micronucleate cells number in buccal mucosa of cancer patients when compared with healthy control subjects [86]. Patients with lung cancer presented the highest number of buccal cells with MN per 1000 cells [86]. A study from Bonassi and coworkers under the frame of the HUman MicroNucleus project on eXfoLiated buccal cells ($HUMN_{XL}$) was conducted combining data of buccal MN values of 5424 subjects from 30 laboratories with the aim of studying several conditions that could impact MN frequency [87]. Significant increases in MN frequency were observed for the majority of occupational exposures and also cancer diagnosis. Associations between elevated MN frequencies was observed for oropharyngeal cancer, respiratory cancer and all the other cancers pooled together. Also, in this study no associations were found for elevated MN frequency and neurodegenerative diseases [87]. A subsequent systematic review and meta-analysis from Bolognesi and coworkers focusing in the clinical application of the micronucleus test in exfoliated buccal cells suggests that the MN assay can be useful in the prescreening and follow-up of precancerous oral lesions [83]. Increased MN frequency was observed in oral and neck cancer patients and leukoplakia, a premalignant stage of cancer, when compared with controls. Besides, in this study, an increased

MN frequency was observed for patients with Alzheimer's disease and Down syndrome [83].

The MN test in urothelial cells has also been used to address the possibility of the assay to be conducted as a diagnostic tool. An increased MN frequency in urothelial cells has been observed in cancer patients [88-90]. Arora and coworkers performed a study with 30 cases of urine samples, where 15 were reported as normal urothelial cells and the other 15 as atypical urothelial cells that later turn out as malignant. MN were only observed in atypical cells (2.53 ± 0.99%), whereas no MN was found in the control cases [88]. A study using MN test in urothelial cells was conducted in women diagnosed with cervix cancer. This study reported an elevated frequency of micronucleated cells in 72% of the cancer patients when compared to 16.7% of controls [89]. Another study observed an increased MN frequency in cervix cancer patients, and found a linear association between the mean MN count and the stage of cervix cancer [90].

These studies suggest that the MN test in exfoliated cells might be an important indicator for malignancy and other diseases. The use of this type of cells may be advantageous in screening programs for occupational and environmental exposures and also for other human diseases, especially given the non-invasive method for collecting the samples.

## CONCLUSION

We are all exposed to several harmful agents in our daily life activities, being those exposures possibly responsible for health impairments in the future. Research focused on the development of integrated approaches regarding the health risk assessment of populations exposed to hazard compounds is of paramount importance. Thus, genotoxicity biomarkers used in environmental and occupational health have been increasingly applied in human population studies.

Cytogenetic biomarkers are the most frequently used endpoints given their sensitivity to measure exposures of genotoxic agents and their role as early cancer risk predictors. These biomarkers are usually measured in lymphocytes, as the cytogenetic damage present in PBLs reflects cumulative exposure events and possible health consequences, as carcinogenesis, which can be related to chronic genotoxic exposure. CAs are the most validated biomarker in this regard although time-consuming. On the other hand, MN assay is a very promising alternative that is more appropriate for human biomonitoring, since it offers an easier technical procedure than SCE and CA assays, and has also been widely associated with cancer risk predictivity in several studies.

In order to efficiently use these cytogenetic endpoints in human biomonitoring is important to be aware of the need for the biomarkers to be validated and to have good protocols and sufficient technical expertise in the laboratories that conduct the assays.

One of main issues concerning cytogenetic biomarkers is the interpretation of altered levels of these indicators at individual level. Traditionally, risk predictions are only valid at a group level and therefore effect of inter-individual variability is removed. Hence, it will be important to conduct further research regarding the alterations on these endpoints at individual level.

Cytogenetic biomarkers such as SCEs, CAs and MN are extensively used as early predictors of clinical disease that can contribute to the implementation of new effective disease prevention policies in occupational and environmental settings. Further human biomonitoring studies encompassing cohorts with larger populations should continue to be conducted with the aim of obtaining a possible cut-off/control level that might be indicative of cancer/exposure risk prediction.

## ACKNOWLEDGMENTS

The microscopic photos were kindly provided by Solange Costa (for Figures 2, 4 and 5) and Fátima Brandão (for Figure 3). Figure 4 was kindly provided by Solange Costa. The work developed by Armanda Teixeira-Gomes and Solange Costa is supported by the Portuguese Foundation for

Science and Technology (FCT) under the grants SFRH/BD/121802/2016 and SFRH/BPD/100948/2014, respectively.

## REFERENCES

[1] IPCS. Environmental Health Criteria 155: *Biomarkers and risk assessment: Concepts and Principles*. World Health Organization, Geneva, Switzerland., 1993.

[2] WHO. *Human biomonitoring: facts and figures*. World Health Organization Regional Office for Europe, Copenhagen., 2015.

[3] Angerer, J; Ewers, U; Wilhelm, M. Human biomonitoring: state of the art. *Int J Hyg Environ Health.*, 2007, 210(3-4), 201-28.

[4] IPCS. *Concise international chemical assessment document (CICAD); no. 40 - Formaldehyde*. World Health Organization, Geneva, Switzerland., 2002.

[5] Klaassen, CD; Amdur, MO. *Casarett and Doull's toxicology: the basic science of poisons*. McGraw-Hill New York., 1996.

[6] Silins, I; Hogberg, J. Combined toxic exposures and human health: biomarkers of exposure and effect. *Int J Environ Res Public Health.*, 2011, 8(3), 629-47.

[7] Links, JM; Kensler, TW; Groopman, JD. Biomarkers and mechanistic approaches in environmental epidemiology. *Annu Rev Public Health.*, 1995, 16, 83-103.

[8] Kyrtopoulos, SA. Biomarkers in environmental carcinogenesis research: striving for a new momentum. *Toxicol Lett.*, 2006, 162(1), 3-15.

[9] Ladeira, C; Viegas, S. Human Biomonitoring–An overview on biomarkers and their application in Occupational and Environmental Health. *Biomonitoring.*, 2016, 3(1).

[10] Costa, S; Carvalho, S; Costa, C; Coelho, P; Silva, S; Santos, LS; et al. Increased levels of chromosomal aberrations and DNA damage in a group of workers exposed to formaldehyde. *Mutagenesis.*, 2015, 30(4), 463-73.

[11] Costa, S; Ceppi, M; Costa, C; Silva, S; Pereira, C; Laffon, B; et al. The cytokinesis-block micronucleus (CBMN) assay in human populations exposed to styrene: A systematic review and meta-analysis. *Mutat Res*., 2016, 770(Pt A), 92-105.

[12] Costa, S; Coelho, P; Costa, C; Silva, S; Mayan, O; Santos, LS; et al. Genotoxic damage in pathology anatomy laboratory workers exposed to formaldehyde. *Toxicology*., 2008, 252(1-3), 40-8.

[13] Costa, S; Pina, C; Coelho, P; Costa, C; Silva, S; Porto, B; et al. Occupational exposure to formaldehyde: genotoxic risk evaluation by comet assay and micronucleus test using human peripheral lymphocytes. *J Toxicol Environ Health A*., 2011, 74(15-16), 1040-51.

[14] Costa, S; Garcia-Leston, J; Coelho, M; Coelho, P; Costa, C; Silva, S; et al. Cytogenetic and immunological effects associated with occupational formaldehyde exposure. *J Toxicol Environ Health A*., 2013, 76(4-5), 217-29.

[15] Teixeira, JP; Gaspar, J; Coelho, P; Costa, C; Pinho-Silva, S; Costa, S; et al. Cytogenetic and DNA damage on workers exposed to styrene. *Mutagenesis*., 2010, 25(6), 617-21.

[16] Costa, C; Silva, S; Neves, J; Coelho, P; Costa, S; Laffon, B; et al. Micronucleus frequencies in lymphocytes and reticulocytes in a pesticide-exposed population in Portugal. *J Toxicol Environ Health A*., 2011, 74(15-16), 960-70.

[17] Costa C; Costa S; Silva S; Coelho P; Botelho M; Gaspar J; et al. DNA damage and susceptibility assessment in industrial workers exposed to styrene. *J Toxicol Environ Health A*. 2012, 75(13-15), 735-46.

[18] Hagmar, L; Bonassi, S; Stromberg, U; Brogger, A; Knudsen, LE; Norppa, H; et al. Chromosomal aberrations in lymphocytes predict human cancer: a report from the European Study Group on Cytogenetic Biomarkers and Health (ESCH). *Cancer Res*., 1998, 58(18), 4117-21.

[19] Bonassi, S; Znaor, A; Ceppi, M; Lando, C; Chang, WP; Holland, N; et al. An increased micronucleus frequency in peripheral blood

lymphocytes predicts the risk of cancer in humans. *Carcinogenesis.*, 2007, 28(3), 625-31.

[20] Mateuca, RA; Decordier, I; Kirsch-Volders, M. Cytogenetic methods in human biomonitoring: principles and uses. *Methods Mol Biol.*., 2012, 817, 305-34.

[21] Salama, SA; Serrana, M; Au WW. Biomonitoring using accessible human cells for exposure and health risk assessment. *Mutat Res.*, 1999, 436(1), 99-112.

[22] Au, WW. Usefulness of biomarkers in population studies: from exposure to susceptibility and to prediction of cancer. *Int J Hyg Environ Health.*, 2007, 210(3-4), 239-46.

[23] Mateuca, R; Lombaert, N; Aka, PV; Decordier, I; Kirsch-Volders, M. Chromosomal changes: induction, detection methods and applicability in human biomonitoring. *Biochimie.*, 2006, 88(11), 1515-31.

[24] Suspiro, A; Prista, J. Biomarkers of occupational exposure do anticancer agents: a mini review. *Toxicol Lett.*, 2011, 207(1), 42-52.

[25] Obe, G; Pfeiffer, P; Savage, JR; Johannes, C; Goedecke, W; Jeppesen, P; et al., Chromosomal aberrations: formation, identification and distribution. *Mutat Res.*, 2002, 504(1-2), 17-36.

[26] Albertini, RJ; Anderson, D; Douglas, GR; Hagmar, L; Hemminki, K; Merlo, F; et al. IPCS guidelines for the monitoring of genotoxic effects of carcinogens in humans. International Programme on Chemical Safety. *Mutat Res.*, 2000, 463(2), 111-72.

[27] Hagmar, L; Stromberg, U; Bonassi, S; Hansteen, IL; Knudsen, LE; Lindholm, C; et al. Impact of types of lymphocyte chromosomal aberrations on human cancer risk: results from Nordic and Italian cohorts. *Cancer Res.*, 2004, 64(6), 2258-63.

[28] Bonassi, S; Au, WW. Biomarkers in molecular epidemiology studies for health risk prediction. *Mutat Res*, 2002, 511(1), 73-86.

[29] OECD. Test No. 473: *In Vitro Mammalian Chromosomal Aberration Test*: OECD Publishing.

[30] Bourner, RD; Parry, EM; Parry, JM. Chemically induced aneuploidy: investigations into chromosome specific effects in mitosis. *Mutat Res.*, 1998, 404(1-2), 191-7.

[31] Roma-Torres, J; Teixeira, JP; Silva, S; Laffon, B; Cunha, LM; Mendez, J; et al. Evaluation of genotoxicity in a group of workers from a petroleum refinery aromatics plant. *Mutat Res.*, 2006, 604(1-2), 19-27.

[32] Fenech, M; Kirsch-Volders, M; Natarajan, AT; Surralles, J; Crott, JW; Parry, J; et al. Molecular mechanisms of micronucleus, nucleoplasmic bridge and nuclear bud formation in mammalian and human cells. *Mutagenesis.*, 2011, 26(1), 125-32.

[33] Fenech, M. Cytokinesis-block micronucleus assay evolves into a "cytome" assay of chromosomal instability, mitotic dysfunction and cell death. *Mutat Res.*, 2006, 600(1-2), 58-66.

[34] Bolognesi, C; Fenech, M. Micronucleus assay in human cells: lymphocytes and buccal cells. *Methods Mol Biol.*, 2013, 1044, 191-207.

[35] OECD. Test No. 474: *Mammalian Erythrocyte Micronucleus Test*: OECD Publishing.

[36] OECD. Test No. 487: *In Vitro Mammalian Cell Micronucleus Test*: OECD Publishing.

[37] Fenech, M. Important variables that influence base-line micronucleus frequency in cytokinesis-blocked lymphocytes-a biomarker for DNA damage in human populations. *Mutat Res.*, 1998, 404(1-2), 155-65.

[38] Bonassi, S; Fenech, M; Lando, C; Lin, YP; Ceppi, M; Chang, WP; et al. HUman MicroNucleus project: international database comparison for results with the cytokinesis-block micronucleus assay in human lymphocytes: I. Effect of laboratory protocol, scoring criteria, and host factors on the frequency of micronuclei. *Environ Mol Mutagen.*, 2001, 37(1), 31-45.

[39] Fenech, M. The advantages and disadvantages of the cytokinesis-block micronucleus method. *Mutat Res.*, 1997, 392(1-2), 11-8.

[40] Latt, SA; Schreck, RR. Sister chromatid exchange analysis. *Am J Hum Genet.*, 1980, 32(3), 297-313.

[41] Wilson, DM; 3rd. Thompson, LH. Molecular mechanisms of sister-chromatid exchange. *Mutat Res.*, 2007, 616(1-2), 11-23.

[42] Helleday, T. Pathways for mitotic homologous recombination in mammalian cells. *Mutat Res.*, 2003, 532(1-2), 103-15.

[43] da Costa, SCB. *Occupational exposure to formaldehyde-Genotoxic damage and susceptibility evaluation in anatomical pathology laboratory workers*: Universidade do Porto (Portugal), 2013.

[44] Taylor, JH; Woods, PS; Hughes, WL. The organization and duplication of chromosomes as revealed by autoradiographic studies using tritium-labeled thymidine. *Proc Natl Acad Sci U S A.*, 1957, 43(1), 122-8.

[45] Natarajan, A; Tucker, J; Sasaki, M. Monitoring cytogenetic damage *in vivo. In: SCOPE 52 Methods to Assess DNA Damage and Repair: Interspecies Comparisons Scientific committee on problems of the environment international council of scientific unions*, 1994, 52, 95-.

[46] Perry, P; Wolff, S. New Giemsa method for the differential staining of sister chromatids. *Nature.*, 1974, 251, 156.

[47] Teixeira, JP; Gaspar, J; Silva, S; Torres, J; Silva, SN; Azevedo, MC; et al. Occupational exposure to styrene: modulation of cytogenetic damage and levels of urinary metabolites of styrene by polymorphisms in genes CYP2E1, EPHX1, GSTM1, GSTT1 and GSTP1. *Toxicology.*, 2004, 195(2-3), 231-42.

[48] WHO. World Health Organization: *Cancer*. http,//www.who.int/cancer/en/, 2017.

[49] Chaganti, RS; Schonberg, S; German, J. A manifold increase in sister chromatid exchanges in Bloom's syndrome lymphocytes. *Proc Natl Acad Sci U S A.*, 1974, 71(11), 4508-12.

[50] Ellis, NA; Proytcheva, M; Sanz, MM; Ye, TZ; German, J. Transfection of BLM into cultured bloom syndrome cells reduces the sister-chromatid exchange rate toward normal. *Am J Hum Genet.*, 1999, 65(5), 1368-74.

[51] Norppa, H; Bonassi, S; Hansteen, IL; Hagmar, L; Stromberg, U; Rossner, P; et al. Chromosomal aberrations and SCEs as biomarkers of cancer risk. *Mutat Res.*, 2006, 600(1-2), 37-45.

[52] Hagmar, L; Brogger, A; Hansteen, IL; Heim, S; Hogstedt, B; Knudsen, L; et al. Cancer risk in humans predicted by increased levels of chromosomal aberrations in lymphocytes: Nordic study group on the health risk of chromosome damage. *Cancer Res.*, 1994, 54(11), 2919-22.

[53] Bonassi, S; Lando, C; Ceppi, M; Landi, S; Rossi, AM; Barale, R. No association between increased levels of high-frequency sister chromatid exchange cells (HFCs) and the risk of cancer in healthy individuals. *Environ Mol Mutagen.*, 2004, 43(2), 134-6.

[54] Brogger, A; Hagmar, L; Hansteen, IL; Heim, S; Hogstedt, B; Knudsen, L; et al. An inter-Nordic prospective study on cytogenetic endpoints and cancer risk. Nordic Study Group on the Health Risk of Chromosome Damage. *Cancer Genet Cytogenet.*, 1990, 45(1), 85-92.

[55] Bonassi, S; Abbondandolo, A; Camurri, L; Dal Pra, L; De Ferrari, M; Degrassi, F; et al. Are chromosome aberrations in circulating lymphocytes predictive of future cancer onset in humans? Preliminary results of an Italian cohort study. *Cancer Genet Cytogenet.*, 1995, 79(2), 133-5.

[56] Hagmar, L; Bonassi, S; Stromberg, U; Mikoczy, Z; Lando, C; Hansteen, IL; et al. Cancer predictive value of cytogenetic markers used in occupational health surveillance programs: a report from an ongoing study by the European Study Group on Cytogenetic Biomarkers and Health. *Mutat Res.*, 1998, 405(2), 171-8.

[57] Bonassi, S; Hagmar, L; Stromberg, U; Montagud, AH; Tinnerberg, H; Forni, A; et al. Chromosomal aberrations in lymphocytes predict human cancer independently of exposure to carcinogens. European Study Group on Cytogenetic Biomarkers and Health. *Cancer Res.*, 2000, 60(6), 1619-25.

[58] Smerhovsky, Z; Landa, K; Rossner, P; Brabec, M; Zudova, Z; Hola, N; et al. Risk of cancer in an occupationally exposed cohort with increased level of chromosomal aberrations. *Environ Health Perspect.*, 2001, 109(1), 41-5.

[59] Boffetta, P; van der Hel, O; Norppa, H; Fabianova, E; Fucic, A; Gundy, S; et al. Chromosomal aberrations and cancer risk: results of

a cohort study from Central Europe. *Am J Epidemiol.*, 2007, 165(1), 36-43.

[60] Bonassi, S; Norppa, H; Ceppi, M; Stromberg, U; Vermeulen, R; Znaor, A; et al. Chromosomal aberration frequency in lymphocytes predicts the risk of cancer: results from a pooled cohort study of 22 358 subjects in 11 countries. *Carcinogenesis.*, 2008, 29(6), 1178-83.

[61] Vodenkova, S; Polivkova, Z; Musak, L; Smerhovsky, Z; Zoubkova, H; Sytarova, S; et al. Structural chromosomal aberrations as potential risk markers in incident cancer patients. *Mutagenesis.*, 2015, 30(4), 557-63.

[62] Murgia, E; Ballardin, M; Bonassi, S; Rossi, AM; Barale, R. Validation of micronuclei frequency in peripheral blood lymphocytes as early cancer risk biomarker in a nested case-control study. *Mutat Res.*, 2008, 639(1-2), 27-34.

[63] Iarmarcovai, G; Ceppi, M; Botta, A; Orsiere, T; Bonassi, S. Micronuclei frequency in peripheral blood lymphocytes of cancer patients: a meta-analysis. *Mutat Res.*, 2008, 659(3), 274-83.

[64] Maffei, F; Zolezzi Moraga, JM; Angelini, S; Zenesini, C; Musti, M; Festi, D; et al. Micronucleus frequency in human peripheral blood lymphocytes as a biomarker for the early detection of colorectal cancer risk. *Mutagenesis.*, 2014, 29(3), 221-5.

[65] Pardini, B; Viberti, C; Naccarati, A; Allione, A; Oderda, M; Critelli, R; et al. Increased micronucleus frequency in peripheral blood lymphocytes predicts the risk of bladder cancer. *Br J Cancer.*, 2017, 116(2), 202-10.

[66] El-Zein, RA; Schabath, MB; Etzel, CJ; Lopez, MS; Franklin, JD; Spitz, MR. Cytokinesis-blocked micronucleus assay as a novel biomarker for lung cancer risk. *Cancer Res.*, 2006, 66(12), 6449-56.

[67] El-Zein, RA; Lopez, MS; D'Amelio, AM; Jr. Liu, M; Munden, RF; Christiani, D; et al. *The cytokinesis-blocked micronucleus assay as a strong predictor of lung cancer: extension of a lung cancer risk prediction model.* Cancer epidemiology, biomarkers & prevention: a publication of the American Association for Cancer Research,

cosponsored by the American Society of Preventive Oncology., 2014, 23(11), 2462-70.

[68] Fenech, M; Holland, N; Chang, WP; Zeiger, E; Bonassi, S. The HUman MicroNucleus Project--An international collaborative study on the use of the micronucleus technique for measuring DNA damage in humans. *Mutat Res.*, 1999, 428(1-2), 271-83.

[69] Stich, HF; San, RH; Rosin, MP. Adaptation of the DNA-repair and micronucleus tests to human cell suspensions and exfoliated cells. *Ann N Y Acad Sci.*, 1983, 407, 93-105.

[70] Suruda, A; Schulte, P; Boeniger, M; Hayes, RB; Livingston, GK; Steenland, K; et al. Cytogenetic effects of formaldehyde exposure in students of mortuary science. *Cancer Epidemiol Biomarkers Prev.*, 1993, 2(5), 453-60.

[71] Viegas, S; Ladeira, C; Nunes, C; Malta-Vacas, J; Gomes, M; Brito, M; et al. Genotoxic effects in occupational exposure to formaldehyde: A study in anatomy and pathology laboratories and formaldehyde-resins production. *J Occup Med Toxicol.*, 2010, 5(1), 25.

[72] Costa, S; Brandão, F; Coelho, M; Costa, C; Coelho, P; Silva, S; et al. Micronucleus frequencies in lymphocytes and buccal cells in formaldehyde exposed workers. *WIT Transactions on Biomedicine and Health.*, 2013, 16, 83-94.

[73] Ying, CJ; Yan, WS; Zhao, MY; Ye, XL; Xie, H; Yin, SY; et al. Micronuclei in nasal mucosa, oral mucosa and lymphocytes in students exposed to formaldehyde vapor in anatomy class. *Biomed Environ Sci.*, 1997, 10(4), 451-5.

[74] Ballarin, C; Sarto, F; Giacomelli, L; Bartolucci, GB; Clonfero, E. Micronucleated cells in nasal mucosa of formaldehyde-exposed workers. *Mutat Res.*, 1992, 280(1), 1-7.

[75] Gonsebatt, ME; Vega, L; Salazar, AM; Montero, R; Guzman, P; Blas, J; et al. Cytogenetic effects in human exposure to arsenic. *Mutat Res.*, 1997, 386(3), 219-28.

[76] Warner, ML; Moore, LE; Smith, MT; Kalman, DA; Fanning, E; Smith, AH. Increased micronuclei in exfoliated bladder cells of

individuals who chronically ingest arsenic-contaminated water in Nevada. *Cancer Epidemiol Biomarkers Prev.*, 1994, 3(7), 583-90.

[77] Moore, LE; Smith, AH; Hopenhayn-Rich, C; Biggs, ML; Kalman, DA; Smith, MT. Micronuclei in exfoliated bladder cells among individuals chronically exposed to arsenic in drinking water. *Cancer Epidemiol Biomarkers Prev.*, 1997, 6(1), 31-6.

[78] Nersesyan, A; Muradyan, R; Kundi, M; Knasmueller, S. Impact of smoking on the frequencies of micronuclei and other nuclear abnormalities in exfoliated oral cells: a comparative study with different cigarette types. *Mutagenesis.*, 2011, 26(2), 295-301.

[79] Piyathilake, CJ; Macaluso, M; Hine, RJ; Vinter, DW; Richards, EW; Krumdieck, CL. Cigarette smoking, intracellular vitamin deficiency, and occurrence of micronuclei in epithelial cells of the buccal mucosa. *Cancer Epidemiol Biomarkers Prev.*, 1995, 4(7), 751-8.

[80] Burgaz, S; Iscan, A; Buyukbingol, ZK; Bozkurt, A; Karakaya, AE. Evaluation of micronuclei in exfoliated urothelial cells and urinary thioether excretion of smokers. *Mutat Res.*, 1995, 335(2), 163-9.

[81] Lehucher-Michel, MP; Di Giorgio, C; Amara, YA; Laget, M; Botta, A. The micronucleus assay in human exfoliated urothelial cells, effect of smoking. *Mutagenesis.*, 1995, 10(4), 329-32.

[82] Reali, D; Di Marino, F; Bahramandpour, S; Carducci, A; Barale, R; Loprieno, N. Micronuclei in exfoliated urothelial cells and urine mutagenicity in smokers. *Mutat Res.*, 1987, 192(2), 145-9.

[83] Bolognesi, C; Bonassi, S; Knasmueller, S; Fenech, M; Bruzzone, M; Lando, C; et al. Clinical application of micronucleus test in exfoliated buccal cells: A systematic review and meta analysis. *Mutat Res Rev Mutat Res.*, 2015, 766, 20-31.

[84] Bloching, M; Hofmann, A; Lautenschlager, C; Berghaus, A; Grummt, T. Exfoliative cytology of normal buccal mucosa to predict the relative risk of cancer in the upper aerodigestive tract using the MN-assay. *Oral Oncol.*, 2000, 36(6), 550-5.

[85] Nersesyan, AK; Adamyan, RT. Micronuclei level in exfoliated buccal mucosa cells of patients with benign and malignant tumors of

female reproductive organs and breast. *Tsitol Genet.*, 2004, 38(3), 72-5.

[86] Yildirim, IH; Yesilada, E; Yologlu, S. Micronucleus frequency in peripheral blood lymphocytes and exfoliated buccal cells of untreated cancer patients. *Genetika.*, 2006, 42(5), 705-10.

[87] Bonassi, S; Coskun, E; Ceppi, M; Lando, C; Bolognesi, C; Burgaz, S; et al. The HUman MicroNucleus project on eXfoLiated buccal cells (HUMN(XL)), the role of life-style, host factors, occupational exposures, health status, and assay protocol. *Mutat Res.*, 2011, 728(3), 88-97.

[88] Arora, SK; Dey, P; Saikia, UN. Micronucleus in atypical urothelial cells. *Diagn Cytopathol.*, 2010, 38(11), 811-3.

[89] Gandhi, G; Sharma, P. The micronucleus test in urothelial cells of cervix cancer patients. *Indian J Hum Genet.*, 2002.

[90] Sundararajan, SK; Natarajan, PS. Kanchana. Micronucleus Assay in Urothelial Cells in Cancer Cervix. *J Clin Diagn Res.*, 2017, 11(3), Xc01-xc3.

In: An Essential Guide to Cytogenetics ISBN: 978-1-53613-370-7
Editors: N. Norris and C. Miller 

*Chapter 3*

# COMPLEX KARYOTYPES AND THEIR DETECTION IN HEMATOLOGIC MALIGNANCIES: THE ONGOING ROLE FOR CLASSICAL CYTOGENETICS

***Helena Urbankova*[*], *PhD***
Department of Hemato-Oncology, Faculty of Medicine and Dentistry, Palacky University Olomouc and University Hospital Olomouc, Olomouc, Czech Republic

## ABSTRACT

The great era for classical cytogenetics started sixty years ago with the description of the twenty-three pairs of human chromosomes (Tjio and Levan, 1956) and the discovery of the Philadelphia chromosome, the first known chromosomal defect associated with a specific type of cancer (Nowell and Hungerford, 1960). Since then many recurrent chromosomal aberrations linked to specific hematological malignancies have been detected. The vast majority of these abnormalities can be detected by modern molecular genetic methods so it might seem there is no longer a need for microscopy. However, there is still a significant portion of

[*] Corresponding Author Email: Helena.Urbankova@fnol.cz.

hemato-oncologic patients for which the classical cytogenetic investigation is appropriate, i.e., cases with complex karyotypes. Complex karyotypes are characterized by the presence of three or more unrelated chromosomal aberrations co-existing in a single clone. Their occurrence is associated with adverse outcomes across the entire spectrum of hematologic malignancies. These aberrations are also a powerful diagnostic indicator for molecular targeted therapies, allogeneic stem cell transplantation or other generally more aggressive treatment strategies. Since the presence of complex karyotypes would be missed when using only molecular genetic methods, this highlights the irreplaceable role of classical cytogenetics as a first-tier analysis for the evaluation of complex structural chromosomal abnormalities in hemato-oncologic patients. Ideally, classical cytogenetics is then followed by more precise molecular genetic methods to identify specific chromosomal aberrations more deeply. Here we focus on the complex karyotype issues in the myelodysplastic diseases, leukemias, lymphomas and multiple myelomas as seen daily in our center.

**Keywords**: complex karyotype, cytogenetics, leukemia, lymphoma

# INTRODUCTION

## Genomic Instability and Chromosomal Aberrations

Genomic instability and the creation of chromosomal aberrations is a hallmark of many types of cancer, including hematologic malignancies. Such genetic instability results in the deregulation of basic cell functions such as proliferation, growth, cell death, replication, angiogenesis, etc. (Hanahan & Weinberg, 2011). The molecular cause of genomic instability in hematologic neoplasms remains unclear, but according to recent sequencing studies, the so called mutator hypothesis seems to be flawed. Instead the sequencing results suggest that the DNA repair genes are frequently free of mutations before therapy. More likely, the mutation status of several other genes appears to be important: the tumor suppressor gene *TP53*, the ataxia telangiectasia mutated (*ATM*) gene and the cyclin-dependent kinase inhibitor 2A gene (*CDKN2A*). Mutations in these genes play a key role in the oncogene-induced DNA replication stress model, the currently most trusted theory of oncogene-induced DNA instability (Gaillard, Garcia-Muse, & Aguilera, 2015; Negrini, Gorgoulis, & Halazonetis, 2010). It is believed that the activation of oncogenes causes DNA replication stress and formation of DNA double strand breaks (DSBs) in precancerous lesions. The continuing generation of DSBs leads to genomic instability and together with defects in the DNA repair and checkpoint genes (*ATM* and *TP53*) results in cancer (Bartkova et al., 2005; Halazonetis, Gorgoulis, & Bartek, 2008). Nevertheless, the precise nature of oncogene-induced replicative stress is still not clear.

In the last 5 years, the deep sequencing of the cancer genomes revealed novel catastrophic mechanisms as the cause of complex chromosomal changes, i.e., chromothripsis, kataegis and chromoplexy. One or more chromosomes carry multiple and adjacent intrachromosomal rearrangements. Such aberrant chromosomes underwent deep fragmentation and further resealing and are no longer regarded as an independent item. More probably this deep fragmentation occurred during a unique catastrophic event (Palumbo & Russo, 2016; Stephens et al., 2011). Interestingly, this hypothesis contrasts starkly with previously described models of carcinogenesis as the consecutive accumulation of aberrations. Evidence for chromothripsis is found in different types of cancer, e.g., multiple myeloma, medulloblastoma, neuroblastoma and colorectal cancers. Moreover, in some studies, chromothripsis was associated with more aggressive types of cancer. This discrepancy needs to be further examined.

Generally, genomic instability generates point mutations leading to large chromosomal rearrangements and possibly results in complex changes affecting many chromosomes. As tumorigenesis is still believed to be a multistep process, we can distinguish between primary and secondary abnormalities, which are changing and accumulating during cancer development (Heim & Mitelman, 1989).

Primary aberrations are connected with the earliest stages of cancer and are mostly found as sole karyotypic changes associated with the already described molecular consequence, e.g., activation of a particular oncogene or inactivation of a specific tumor-suppressor gene. However sometimes primary aberrations can present simply as submicroscopic mutations and may not be visible in the karyotype.

Secondary aberrations develop in cells with existing primary changes during later stages of the disease, probably as the consequence of the altered cancer cell metabolism (e.g., oxidative stress, defect checkpoints, mutations in DNA repair genes, etc.). Secondary aberrations are generally more numerous and less specific, but we still see non-randomness in many of them. It remains to be determined if all of them are pathologically relevant and have functional impacts on the course of the disease.

There are specific changes with known impacts on the course of the disease, but there is also a wide spectrum of aberrations with unclear functions. Elucidating this issue would require both an overview of all chromosomal aberrations that occur in cancer genomes as well as their analyses in large cohorts of patients. Important information can be lost when using only molecular genetic analyses, which either focus on particular targets and can therefore miss aberrations in other locations, or do not distinguish between different clones in a population of tumor cells. It is still uncertain whether the proposed scenario of primary and secondary changes is correct. It is well known that a primary change detected in one type of neoplasm can also occur as a secondary change in a different neoplasm and have a different clinical impact. Moreover, there is one more criterion: the aberrations must be clonal, meaning that they must be detected in at least two identical cells and, in the case of a loss of genetic material, even three cells. The significance of multiple chromosomal changes remains unexplained. Despite of this it is now clear that patients diagnosed with a hematologic malignancy that have three or more aberrations in the same clone (the so called complex karyotypes) are associated with adverse outcomes.

There are two types of chromosomal aberrations - structural and numerical. They lead to balanced/unbalanced relocations of chromosomal material or result in a loss or gain of entire chromosomes or chromosomal segments. Based on the type of aberration present, different detection methods are available.

## Methods of Detection of Chromosomal Aberrations

Conventional cytogenetics is based on the chromosomes present in the metaphase stage of the cell cycle, when chromatin is highly condensed and a typical chromosome morphology is visible. Therefore cells (mostly from the peripheral blood, bone marrow or other biologic material) are grown short term in a culturing medium and arrested in a metaphase using a mitotic inhibitor. The cell suspension is further harvested and fixed using Carnoy's solution. The most frequently used technique to stain chromosome bands in hemato-oncology is G-banding. The cell suspension is pretreated with trypsin and stained with a Giemsa staining solution. Twenty-three chromosome pairs can then be identified in humans based on their size, position of a centromere and a specific bright and dark banding pattern. The assembled karyotype is described using the International System for Human Cytogenetic Nomenclature (ISCN) 2016 (McGowan-Jordan J., Simons A., & Schmid M., 2016) and the detected abnormalities are defined. The advantage of this technique is that it allows the identification of balanced/unbalanced aberrations in a single cell at a low cost. Unfortunately, it has a relatively low resolution depending on the number and type of dividing cells and the quality of metaphase chromosomes, which can allow smaller aberrations to escape observation. Therefore, molecular cytogenetic methods based on complementary hybridization of a fluorescent labeled probe to patient`s DNA were introduced.

These techniques brought a rapid improvement to the detection of chromosomal rearrangements. Fluorescence *in situ* hybridization (FISH) is the most frequently used technique. It was developed in 1982 by Langer-Safer et al., (Langer-Safer, Levine, & Ward, 1982) and its use for chromosome classification was described later in 1986 by Pinkel et al., (Pinkel, Straume, & Gray, 1986). FISH is able to detect and localize the presence or absence of a specific DNA sequence either on a particular chromosome or in an interphase nuclei without the need of metaphase chromosomes.

A large number of fluorescent probes targeting various sequences of the genome are now commercially available. FISH probes can be divided into three categories: locus specific, centromeric and whole chromosome painting probes. The selection of the appropriate probe for the detection of a specific chromosomal rearrangement associated with a particular type of hemato-oncologic disease depends on previous knowledge of the possible aberration. We can only routinely use a combination of two or three fluorochromes in a single hybridization and are therefore able to detect only a limited number of changes. Performing multiple hybridizations is time consuming and expensive. For these reasons using the FISH technique alone is not ideal for detecting multiple chromosomal aberrations and complex karyotypes. Similarly, the use of a singular molecular method such as allele-specific polymerase chain reaction (PCR), multiplex ligation-dependent probe amplification (MLPA), Sanger sequencing, etc., is also inadequate.

The multicolor FISH (M-FISH) approach is a much better tool. It was developed in the 1990s and published by Schröck et al., (Schrock et al., 1996) and Speicher et al., (Speicher, Gwyn Ballard, & Ward, 1996). This method is based on the complementary hybridization of probes to chromosomes in metaphase. The probes are labelled with a different fluorochrome combination specific to each chromosome. This technique allows the rapid and unequivocal detection of complex chromosomal rearrangements, whether they are balanced or unbalanced. Unfortunately the resolution of the M-FISH approach is also limited because of its correlation to chromosomes.

The introduction of a comparative genomic hybridization (CGH) was a key innovation. CGH is based on a competitive hybridization between the reference and tumor genomic DNA to the normal metaphase chromosomes. This approach could reveal unbalanced chromosomal aberrations by a single genome wide experiment (Kallioniemi et al., 1992). However, this technique also has disadvantages: its inability to detect balanced changes of genomes together with the relatively low resolution caused by the hybridization to chromosomes.

A few years later a variant of this technique was introduced: an arrayCGH (Pinkel et al., 1998; Solinas-Toldo et al., 1997). This method is also based on a competitive hybridization of the reference and tumor genomic DNA, but instead of binding to chromosomes it is applied to a chip with a spectrum of immobilized fragments of DNA. This enables the arrayCGH technique to detect genome wide unbalanced abnormalities with a high resolution defined by size and genome localization of the used fragments. ArrayCGH became a fast and reliable diagnostic tool for many cancer types, as well as hemato-oncologic diseases (Carter, Fiegler, & Piper, 2002; Fiegler et al., 2003; Vermeesch et al., 2007; Vermeesch et al., 2005). When combined with conventional cytogenetics and FISH, it provides a powerful methodology for the detection of complex chromosomal aberrations in hematologic malignancies (Knijnenburg et al., 2005; Kolialexi, Tsangaris, Kitsiou, Kanavakis, & Mavrou, 2005).

Next generation sequencing (NGS) strategies were developed recently (Sikkema-Raddatz et al., 2013). They are able to detect either balanced or unbalanced aberrations through the whole genome, but they have not added any major findings to the cytogenetic landscape of hematologic neoplasms. Moreover, they are quite expensive and require extensive biostatistical analysis. This is most likely the reason why these genome wide sequencing strategies are not currently widely used in diagnostic practice.

As it is well known that complex chromosomal changes impact prognosis in many hemato-oncologic diseases including the myelodysplastic syndrome (MDS), acute myeloid leukemia (AML), B-cell lymphoblastic leukemia/lymphoma (B-ALL), chronic myeloid leukemia (CML), chronic lymphocytic leukemia (CLL) and multiple myeloma (MM).

Unfortunately there is still a certain bias in defining a complex karyotype and its impact in various hematologic malignancies and there is a need for standardization (Peterson, 2017).

# Complex Karyotype in Selected Hematologic Malignancies

## Myelodysplastic Syndrome (MDS)

Myelodysplastic syndrome is a heterogeneous group of clonal bone marrow disorders characterized by impaired maturation of the hematopoietic cells associated with one or more peripheral blood cytopenias, which can also progress to acute myeloid leukemia (Vardiman, 2003). The current diagnostics follow the established WHO (World Health Organization) 2016 criteria based on histology, blast counts and cytogenetic findings (Arber et al., 2016). At the time of diagnosis, recurrent chromosomal aberrations are found in 40-70% of patients with primary MDS and these aberrations predict the rate of survival and the risk of leukemic transformation (Haase et al., 2007). The majority of these chromosomal changes are unbalanced (exhibiting loss of a part or an entire chromosome), but balanced translocations and complex derivative chromosomes can also be detected. The most frequent cytogenetic aberrations include: del(5)(q) found in 10-20% of primary MDS, -7/del(7)(q) (10%), +8 (10%), del(20)(q) (5%) and others (Olney & Le Beau, 2001). All of these changes could be seen as solo aberrations or as a part of a complex karyotype, which has a big impact on the prognosis and the course of the disease. For example patients with isolated del(5)(q) have a good prognosis with a median survival of 4.8 years, whereas patients carrying the del(5)(q) as a part of a complex karyotype have a poor prognosis with early progression to AML. Complex karyotypes are observed in 10% of primary MDS and up to 90% of therapy-related MDS (Schanz et al., 2012). According to a Revised International Prognostic Scoring System for myelodysplastic syndromes (IPSS-R) (Greenberg et al., 2012), a complex karyotype in MDS is defined by the presence of three chromosomal changes in the poor prognostic group (with median survival 1.5 years), whereas the very poor prognostic group carry more abnormalities (median survival 0.7 years). Moreover according to the study of Göhring G. et al., (Gohring et al., 2010) aimed at the identification of prognostic factors in childhood myelodysplastic syndrome, it is important to define even the types of aberrations included in a complex karyotype. They defined the so called structurally complex karyotype, which is characterized by three or more chromosomal aberrations, at least one of which is structural. Patients with complex karyotypes but without structural rearrangement had a five-year overall survival similar to patients with normal karyotype, while patients with a structurally complex karyotype had a very poor outcome. Other authors studied the so called monosomal karyotype, which is defined by the presence of two or more autosomal monosomies or one monosomy with at least one structural abnormality. Such monosomal karyotypes seem to be associated with poor risk or very poor risk IPSS-R categories, but its prognostic value remains controversial. Other studies show that the impact of a complex karyotype as detected by conventional cytogenetics is increased by the presence of cytogenetically invisible mutations in the *TP53* gene, which can be detected by sequencing analyses. Patients with the *TP53* mutations and complex karyotype were associated with an extremely poor survival rate and a frequent early relapse, whereas outcomes were unambiguously better in the *TP53*-mutated patients without complex karyotype (Yoshizato et al., 2017).

In summary, not only is the presence of complex karyotype significant, but the number and the type of changes are also important considerations for establishing the diagnosis, prognosis and therapeutic strategies in patients with MDS. For these reasons the role of conventional cytogenetics remains pivotal in MDS. Cytogenetic results are very often complicated and biased, and therefore must be evaluated very carefully by a skilled cytogeneticist or a clinician experienced in cytogenetics in order to correctly interpret the karyotypic data. The combination of conventional cytogenetics with the results of molecular cytogenetic and genomic methods can produce more refined results and improve the risk stratification of MDS patients.

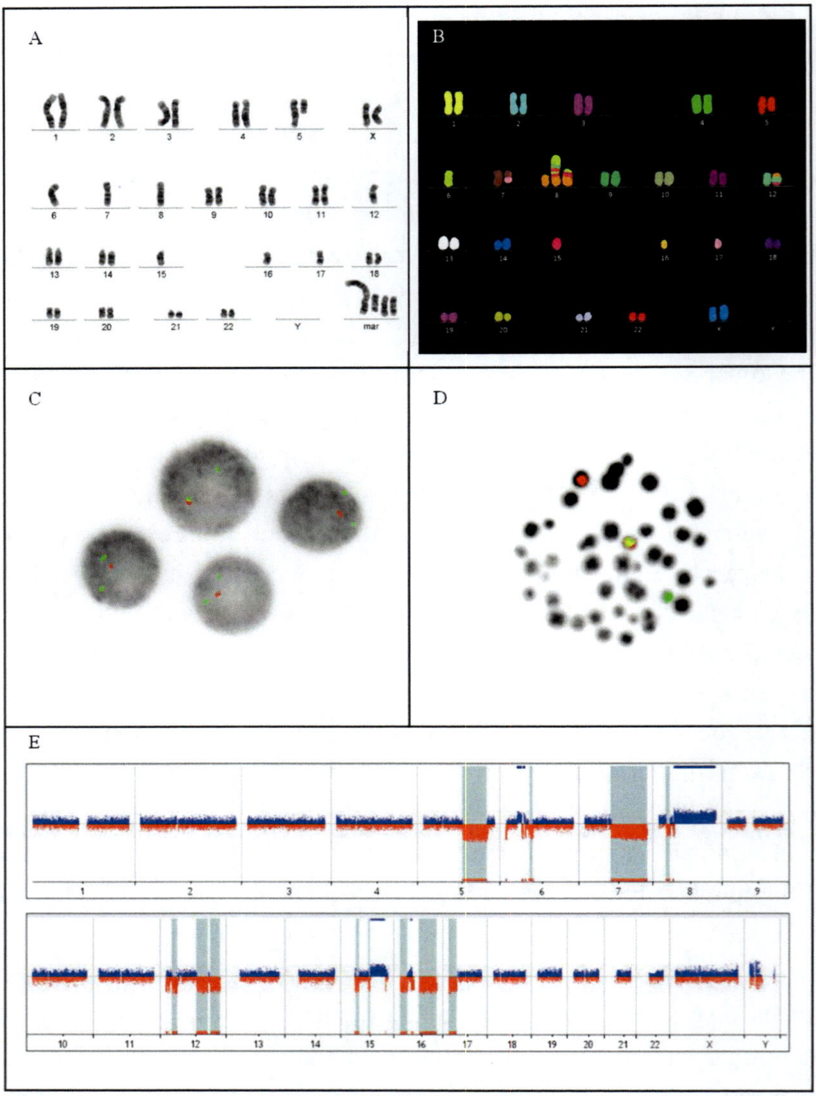


All images were created in the Laboratory of Cytogenetics and Molecular Cytogenetics, Dpt. of Hemato-oncology Olomouc, Czech Republic, namely by: Kropackova J., Holzerova M. and Urbankova H.

Figure 1. Results of cytogenetics and molecular cytogenetic/genetic methods used to determine complex genomic aberrations in a patient with MDS diagnosed in our center.

A. The complex karyotype detected by conventional cytogenetics (G-banding) of the bone marrow sample, marker chromosomes with unknown origin are present. The karyotype was described according to ISCN 2016: 43,XX,del(5)(q?),-6,-7,-8,-12,-15,-16,-17,?der(18),+4mar[cp24].
B. By M-FISH we proved the presence of the complex karyotype including both structural and numerical aberrations and we were able to specify more precisely particular abnormalities and marker chromosomes, e.g., monosomy 6, derivative chromosome der (7)t(7;17), trisomy 8 with two derivative chromosomes 8, etc. The karyotype was corrected: 43,XX,del(5)(q31q?35), -6,?der(7)t(7;17)(p?12;q?12),der(8)t(6;8)(?p;p?12)ins(8;15)(p?12;q?),der(<u>8</u>)(6qter›6q?13::12?›12?::8?›8?::15q?›15q?::6?›6?::15q?›15q?::8p11›8qter),+8,der(12)t(8;12)(?q;p?12)t(12;15)(q?24;q?),-15,-16,-17[10]
C. Interphase FISH with the locus specific probe XL Del(5)(q31) (MetaSystems, Germany) confirmed the deletion of 5q31 in 81% of cells. The orange labeled probe is designed to hybridize to the EGR1 locus at 5q31, whereas the green probe covers specific locus at 5p15 and functions as a control probe.
D. The result of metaphase FISH with centromeric probes specific for the centromeric regions of chromosome 7 (labeled in orange) and 17 (green) (Abbott Molecular, USA). The yellow signal is proving the presence of the dicentric chromosome dic(7;17). Based on this finding, which was not visible previously, the original description of der(7)t(7;17) was corrected for dic(7;17)(q11;p11) with consequently demonstrated deletion of the *TP53* gene.
E. ArrayCGH allowed us to detect already known changes as well as other previously hidden abnormalities and to precisely define the breakpoint regions and the aberrations of small sizes. In total we found 26 unbalanced genomic changes through the entire genome (Sure Print G3 ISCA CGH+SNP Microarray, 4x180K, Agilent), 21 losses (5q21.1q34, 6p25.3, 6p21.1, 6p12.3, 6p11.2q12, 6q12, 7q11.22q36.3, 8p21.3p12, 8p12p11.21, 12p13.33, 12p13.31, 12p12.3p11.22, 12q21.2q23.3, 12q24.11q24.33, 15q11.2q13.2,

15q21.3, 15q26.2, 16p13.3p12.3, 16p12.1, 16p12.1q24.3 and 17p13.3p11.2) and 5 gains (6p22.1p21.1, 6p21.1p12.3, 8p11.21q24.3, 15q21.3q26.2 and 16p12.1). Chromosomes are listed from the left to the right, from 1 to 9 and 10 to X and Y, respectively. The regions of gain are represented by the blue bars along the chromosomes, the regions of loss by the red ones.

## Adult Acute Myeloid Leukemia (AML)

Acute myeloid leukemia is a neoplasm of the myeloid line of blood cells characterized by an accumulation of immature myeloid blasts in the bone marrow. This leads to an abnormal hematopoesis and to a lack of differentiated granulocytes, monocytes, thrombocytes or erythrocytes. Based on the morphology, cytochemistry, immunophenotyping, genetic features and clinical characteristics, subgroups of AML can be defined with differing clinical relevance. There are currently seven subgroups of AML described (Arber et al., 2016; Swerdlow S. H., Campo E., Harris N. L., al., & 2008) and the majority of them has detectable clonal chromosomal aberrations. These aberrations, either numerical or structural, are summarized at the Mitelman Database of Chromosome Aberrations and Gene Fusions in Cancer (Mitelman, 2017). Cytogenetics provides strong prognostic information for prediction of the time of remission and post-remission therapy. Chromosomal abnormalities with good prognoses include t(8;21)(q22;q22.1), inv(16)(p13.1q22) or t(16;16)(p13.1;q22) and t(15;17)(q22;q21). Patients carrying a normal karyotype are placed in the AML category with an average risk and require further molecular genetic analyses for accurate risk stratification. Patients with detected myelodysplasia-related changes del(5)(q), monosomies of chromosomes 5 or 7, translocations or inversions of chromosome 3, t(6;9), t(9;22), or abnormalities of chromosome 11q23 (*MLL* gene) have a particularly poor prognosis. Despite the fact that each specific category of AML has its own prognostic and treatment implications, for practical purposes antileukemic therapy at the diagnosis of AML is similar for all subtypes. The term complex karyotype in AML was introduced in the 1980s and at that time it was already observed to be associated with a poor prognosis (Berger et al., 1987; Levin, Le Coniat, Bernheim, & Berger, 1986). It was found that complex karyotypes are frequently characterized by the presence of certain chromosomal aberrations, for example monosomies of chromosome 5 and 7, deletions of 5q, 7q, 12p and 17p, moreover they are often associated with mutations of the *TP53* gene (Bowen et al., 2009).

The currently reported frequency of complex karyotypes with three and more unrelated chromosomal aberrations is 10-40% in all AML patients and increases with age (Stolzel et al., 2016). According to ELN (European Leukemia Net) recommendations, complex karyotype with three and more changes is classified as the adverse risk category. In comparison, the UK national classification system of Medical Research Council (MRC) for AML established the cutoff for adverse prognosis as the presence of at least four or more aberrations, and they consider the karyotype complex if it has at least five aberrations. Moreover, the definition of complex karyotype in AML is further complicated by the nature of its components. Breems et al., (Breems & Lowenberg, 2011) recently investigated the prognostic value of a monosomal karyotype, which is defined as either one autosomal monosomy and the presence of at least one structural aberration, or at least two distinct autosomal monosomies. They proved that the monosomal karyotype has a highly unfavorable impact on prognosis. On the contrary, it is well known that patients with complex karyotypes containing prognostically favorable aberrations, such as t(8;21), inv(16)/t(16;16) or t(15;17), remain in the good prognostic group despite the presence of the complex karyotype (Peterson, 2017).

Nevertheless, in general an increasing number of abnormalities included in complex karyotypes definitely worsens the overall survival of AML patients. Indeed, there is a need for a uniformly defined complex karyotype as an adverse-risk factor between different institutions. This issue should be figured out in the future by a survival analysis of a large multicenter cohort of AML patients with complex aberrant karyotypes.

## Chronic Lymphocytic Leukemia (CLL)

Chronic lymphocytic leukemia is the most common type of leukemia in adults. It affects B-lymphocytes, which grow in an uncontrolled manner and accumulate in the bone marrow, peripheral blood and lymph nodes. In the past the use of conventional cytogenetics in the detection of chromosomal aberrations in CLL was not very successful due to neoplastic CLL cells having a very low mitotic activity, resulting in a lack of metaphase chromosomes in the cultures of peripheral blood cells.

Mitogenic stimulation of the CLL cultured cells with CpG-oligonucleotides and interleukin 2 was identified more than ten years ago as a reliable and reproducible tool that could obtain metaphase chromosomes and enable the assembly of a karyotype, even in CLL samples (Dicker, Schnittger, Haferlach, Kern, & Schoch, 2006). After an appropriate stimulation, chromosomal aberrations can now be detected in more than 80% of patients with CLL. The most frequent changes are deletions of the 13q14, 11q22, 17p13 and trisomy of chromosome 12 (Dohner et al., 2000). Deletions of the 11q22 (*ATM* gene) and 17p13 (*TP53* gene) are considered high risk factors with a poor outcome (Rigolin et al., 2017). The presence of complex aberrant karyotype was found in 10-20% of CLL patients (Haferlach, Dicker, Schnittger, Kern, & Haferlach, 2007; Mayr et al., 2006). Interestingly, many breakpoints are located in regions of recurrent losses in CLL, e.g., 13q14 and 17p13, which means conversely that complex karyotypes in CLL are often associated with 17p abnormalities (either deletions or mutations of the *TP53* gene) (Fink et al., 2006). In addition, the CLL subgroup with a complex karyotype seems to be strongly associated with an unmutated *IGVH* status and CD38 expression (Haferlach et al., 2007). According to recent studies, the complex karyotype became a strong independent predictive marker of rapid progression, resistance to therapy, early relapse and a short overall survival (Kujawski et al., 2008; Mayr et al., 2006; Puiggros et al., 2017; Van Den Neste et al., 2007).

Conventional cytogenetics represents the best tool for the detection of complex karyotypic abnormalities. Nevertheless, we also routinely use an interphase FISH technique which can estimate an aberrations in six precisely selected loci. The technique can provide information even in the absence of metaphase chromosomes or when the targeted abnormality is below the resolution of classical cytogenetics. We use commercially available probes for the numerical aberrations of 6q21, 8q24 (*MYC* gene), 11q22 (*ATM* gene), chromosome 12, 13q14, 13q34 and 17p13 (*TP53* gene) and rearrangement of the *IGH* gene (14q32). To complete the whole picture of a genome we perform an arrayCGH. The final karyotype is based on a combination of the results of all these methods.

Our data on complex karyotypes in CLL patients are in concordance with other authors (Kruzova L., 2017). We have detected complex karyotype in approximately 12% of untreated CLL cases diagnosed in our center and we have proven their adverse impact on the overall survival of patients. We did not observe any worsening effects due to the presence of *ATM* or *TP53* deletions as a part of a complex karyotype, which is in agreement with the finding of Thompson et al., (Thompson et al., 2015) and others. Moreover, we investigated particular chromosomal aberrations included in the complex karyotypes and their impact on the course of the disease in detail, but found no specific influence on the outcome in any way. Also we have suggested that patients with complex karyotypes are recommended for new treatments which are applied already in the first line of therapy.

## Mantle Cell Lymphoma (MCL)

MCL comprises 3-10% of B-cell non-Hodgkin`s lymphoma (B-NHL). Unlike other NHL types, nearly all cases are characterized by some level of the leukemic course of the disease with neoplastic cells detectable in the peripheral blood, as well as in bone marrow and other locations. This is advantageous for conventional cytogenetics, as the MCL infiltrated samples are available by non-invasive approach. The cytogenetic hallmark of the disease is the presence of t(11;14)(q13;q32), first identified by

Banks et al., in 1992 (Banks et al., 1992), which leads to a juxtaposition of the *CCND1* gene to the *IGH* locus and to an overexpression of the cyclin D1. In approximately 20% of patients the translocation is a part of a complex karyotype (Gazzo et al., 2005) with the recurrent loss of der(11) and other changes. Complex karyotype is associated with a more aggressive disease and a shorter overall survival in MCL, but it is not clear if it can be regarded as an independent prognostic factor (Cohen et al., 2015; Sarkozy et al., 2014). This disease is quite rare and the data were mostly retrieved retrospectively. That is why our patients had been treated with different therapies and it is impossible to reliably compare their outcomes. We have analyzed cases diagnosed with MCL in our center from 2006 until now and have found complex karyotypes in 19% of patients (Obr A., unpublished data). Complex karyotypes were frequently accompanied by mutations in the *TP53* gene and a poor prognosis. We would need to analyze a larger cohort of patients treated with similar modern approaches to prove the independent prognostic impact of the presence of a complex karyotype in MCL.

## Multiple Myeloma (MM)

Multiple myeloma is a cancer of the plasma cells - terminally differentiated B-lymphocytes which are normally responsible for the production of antibodies. Neoplastic plasma cells can form a mass in the bone marrow or soft tissue. Multiple myeloma is diagnosed based on the detection of abnormal levels of antibodies in the peripheral blood or urine, the finding of neoplastic plasma cells in bone marrow biopsy, and bone lesions revealed by medical imaging. Cytogenetic analysis is an important part of MM diagnostics as chromosomal aberrations are powerful indicators of prognosis. Patients can be stratified into two prognostic categories according to the chromosome ploidy status and the rearrangements of the *IGH* gene. Generally, the hypodiploid group with t(4;14)(p16;q32) or t(14;16)(q32;q23) is associated with a high-risk disease, whereas the hyperdiploid patients with t(11;14)(q13;q32) are considered a better prognostic group. During the disease progression the secondary chromosomal aberrations are developing. The most frequent are *MYC* rearrangements, deletions of the 13q, 17p, deletions of the 1p and/or amplification of the 1q. These secondary changes are characteristic of a highly proliferating and rapidly developing disease (Sawyer, 2011). Nevertheless, conventional cytogenetics is difficult because of the low level of bone marrow infiltration and the low mitotic index of plasma cells. Therefore cytogenetics is either unsuccessful or the results show normal karyotype in a majority of cases. In cases where abnormal karyotype is identified in a highly proliferative disease frequently multiple structural and numerical changes are found. Moreover patients often have various abnormal clones. Interphase FISH detects numerical and structural abnormalities in up to 90% of patients (Drach et al., 1995). To deal with a low infiltration of plasma cells in bone marrow samples, conventional interphase FISH was adapted to detect changes specifically in the plasma cells using a cytoplasmic light chain immunofluorescent labeling or purified CD138+ cells (Ahmann, 1998). The disadvantages of this approach are the restricted number of analyzed loci and lack of metaphase chromosomes, which subsequently also limits the detection of the complex karyotypes in MM. According to Nemec et al., (Nemec et al., 2012), metaphase cytogenetic analysis revealed the presence of complex karyotype in 19% of cases. Patients with a complex karyotype, translocation t(4;14) and/or gain of the 1q21 locus had a shorter time to progression and a lower overall survival.

Recently, the arrayCGH technique has also found its place in the routine analysis of the genomic profile of the plasma cell, as it uses DNA from separated CD138+ cells. This method is able to identify multiple novel genetic aberrations that previously escaped detection. Nevertheless, the precise definition of the complex karyotype could not be matched in the majority of MM cases as we have no cytogenetic data proving a co-existence of various genomic changes in one clone. Perhaps this is the reason the complex karyotype is not discussed in MM as often as in other hematologic malignancies, even though it is probably present in a vast majority of them.

## CONCLUSION

There is accumulative evidence that classical cytogenetics remains a valuable basic method for the analysis of genetic aberrations in the hemato-oncologic patients over decades. Previously it has been partially replaced by more precise molecular cytogenetics and genomic methods. However, with the recently recognized significance of the complex karyotypic changes in a wide spectrum of hematologic neoplasms, its role is again becoming crucial in the diagnostics and prognosis prediction. The correlation of cytogenetic results with clinical data and modern molecular genetic methods may provide clinicians with a broad and complex overview of biological events taking place in tumor cells and may help them in therapy related decision making. Whereas in some hemato-oncologic diseases the definition of the complex karyotype is clear, in others it remains a topic of discussion. For example, in AML not only the number of unrelated clonal chromosomal aberrations but also its specific composition appear to be important. Complex karyotype is well defined and reported in CLL where it became an independent prognostic marker. On the contrary in other hemato-oncologic entities with problematic neoplastic infiltration of the bone marrow or peripheral blood (e.g., MM, B-/T-cell lymphomas other than MCL etc.), complex karyotypes are difficult to check with only conventional cytogenetics. Regardless of the scarcity of cytogenetic data, the results of other molecular cytogenetic methods show evidence of multiple genomic aberrations. In the future we can probably expect some adjustments to the definitions of certain questionable complex karyotypes in hematologic neoplasms.

In summary, conventional cytogenetics ideally complemented by other molecular cytogenetic and genomic methods remains an essential tool for the detection of the complex karyotype as an indicator of a poor prognosis across many hemato-oncologic diseases.

## References

Arber, D. A., Orazi, A., Hasserjian, R., Thiele, J., Borowitz, M. J., Le Beau, M. M., Vardiman, J. W. (2016). The 2016 revision to the World Health Organization classification of myeloid neoplasms and acute leukemia. *Blood, 127*(20), 2391-2405. doi: 10.1182/blood-2016-03-643544.

Banks, P. M., Chan, J., Cleary, M. L., Delsol, G., De Wolf-Peeters, C., Gatter, K., et al., (1992). Mantle cell lymphoma. A proposal for unification of morphologic, immunologic, and molecular data. *Am J Surg Pathol, 16*(7), 637-640.

Bartkova, J., Horejsi, Z., Koed, K., Kramer, A., Tort, F., Zieger, K., Bartek, J. (2005). DNA damage response as a candidate anti-cancer barrier in early human tumorigenesis. *Nature, 434*(7035), 864-870. doi: 10.1038/nature03482.

Berger, R., Bernheim, A., Ochoa-Noguera, M. E., Daniel, M. T., Valensi, F., Sigaux, F., Boiron, M. (1987). Prognostic significance of chromosomal abnormalities in acute nonlymphocytic leukemia: a study of 343 patients. *Cancer Genet Cytogenet, 28*(2), 293-299.

Bowen, D., Groves, M. J., Burnett, A. K., Patel, Y., Allen, C., Green, C., Linch, D. C. (2009). TP53 gene mutation is frequent in patients with acute myeloid leukemia and complex karyotype, and is associated with very poor prognosis. *Leukemia, 23*(1), 203-206. doi: 10.1038/leu. 2008.173.

Breems, D. A., & Lowenberg, B. (2011). Acute myeloid leukemia with monosomal karyotype at the far end of the unfavorable prognostic spectrum. *Haematologica, 96*(4), 491-493. doi: 10.3324/haematol. 2011.043208.

Carter, N. P., Fiegler, H., & Piper, J. (2002). Comparative analysis of comparative genomic hybridization microarray technologies: report of a workshop sponsored by the Welcome Trust. *Cytometry, 49*(2), 43-48. doi: 10.1002/cyto.10153.

Cohen, J. B., Ruppert, A. S., Heerema, N. A., Andritsos, L. A., Jones, J. A., Porcu, P., ... Blum, K. A. (2015). Complex karyotype is associated with aggressive disease and shortened progression-free survival in patients with newly diagnosed mantle cell lymphoma. *Clin Lymphoma Myeloma Leuk, 15*(5), 278-285 e271. doi: 10.1016/j.clml.2014.12.012.

Dicker, F., Schnittger, S., Haferlach, T., Kern, W., & Schoch, C. (2006). Immunostimulatory oligonucleotide-induced metaphase cytogenetics detect chromosomal aberrations in 80% of CLL patients: A study of 132 CLL cases with correlation to FISH, IgVH status, and CD38 expression. *Blood, 108*(9), 3152-3160. doi: 10.1182/blood-2006-02-005322.

Dohner, H., Stilgenbauer, S., Benner, A., Leupolt, E., Krober, A., Bullinger, L., Lichter, P. (2000). Genomic aberrations and survival in chronic lymphocytic leukemia. *N Engl J Med, 343*(26), 1910-1916. doi: 10.1056/NEJM200012283432602.

Drach, J., Angerler, J., Schuster, J., Rothermundt, C., Thalhammer, R., Haas, O. A., Huber, H. (1995). Interphase fluorescence in situ hybridization identifies chromosomal abnormalities in plasma cells from patients with monoclonal gammopathy of undetermined significance. *Blood, 86*(10), 3915-3921.

Fiegler, H., Carr, P., Douglas, E. J., Burford, D. C., Hunt, S., Scott, C. E., Carter, N. P. (2003). DNA microarrays for comparative genomic hybridization based on DOP-PCR amplification of BAC and PAC clones. *Genes Chromosomes Cancer, 36*(4), 361-374. doi: 10.1002/gcc.10155.

Fink, S. R., Smoley, S. A., Stockero, K. J., Paternoster, S. F., Thorland, E. C., Van Dyke, D. L., Dewald, G. W. (2006). Loss of TP53 is due to rearrangements involving chromosome region 17p10 approximately p12 in chronic lymphocytic leukemia. *Cancer Genet Cytogenet, 167*(2), 177-181. doi: 10.1016/j.cancergencyto.2006.01.005.

Gaillard, H., Garcia-Muse, T., & Aguilera, A. (2015). Replication stress and cancer. *Nat Rev Cancer, 15*(5), 276-289. doi: 10.1038/nrc3916.

Gazzo, S., Felman, P., Berger, F., Salles, G., Magaud, J. P., & Callet-Bauchu, E. (2005). Atypical cytogenetic presentation of t(11;14) in mantle cell lymphoma. *Haematologica, 90*(12), 1708-1709.

Gohring, G., Michalova, K., Beverloo, H. B., Betts, D., Harbott, J., Haas, O. A., Schlegelberger, B. (2010). Complex karyotype newly defined: the strongest prognostic factor in advanced childhood myelodysplastic syndrome. *Blood, 116*(19), 3766-3769. doi: 10.1182/blood-2010-04-280313.

Greenberg, P. L., Tuechler, H., Schanz, J., Sanz, G., Garcia-Manero, G., Sole, F., Haase, D. (2012). Revised international prognostic scoring system for myelodysplastic syndromes. *Blood, 120*(12), 2454-2465. doi: 10.1182/blood-2012-03-420489.

Haase, D., Germing, U., Schanz, J., Pfeilstocker, M., Nosslinger, T., Hildebrandt, B., Steidl, C. (2007). New insights into the prognostic impact of the karyotype in MDS and correlation with subtypes: evidence from a core dataset of 2124 patients. *Blood, 110*(13), 4385-4395. doi: 10.1182/blood-2007-03-082404.

Haferlach, C., Dicker, F., Schnittger, S., Kern, W., & Haferlach, T. (2007). Comprehensive genetic characterization of CLL: a study on 506 cases analyzed with chromosome banding analysis, interphase FISH, IgV(H) status and immunophenotyping. *Leukemia, 21*(12), 2442-2451. doi: 10.1038/sj.leu.2404935.

Halazonetis, T. D., Gorgoulis, V. G., & Bartek, J. (2008). An oncogene-induced DNA damage model for cancer development. *Science, 319*(5868), 1352-1355. doi: 10.1126/science.1140735.

Hanahan, D., & Weinberg, R. A. (2011). Hallmarks of cancer: the next generation. *Cell, 144*(5), 646-674. doi: 10.1016/j.cell.2011.02.013.

Heim, S., & Mitelman, F. (1989). Primary chromosome abnormalities in human neoplasia. *Adv Cancer Res, 52*, 1-43.

Kallioniemi, A., Kallioniemi, O. P., Sudar, D., Rutovitz, D., Gray, J. W., Waldman, F., & Pinkel, D. (1992). Comparative genomic hybridization for molecular cytogenetic analysis of solid tumors. *Science, 258*(5083), 818-821.

Knijnenburg, J., Szuhai, K., Giltay, J., Molenaar, L., Sloos, W., Poot, M., Rosenberg, C. (2005). Insights from genomic microarrays into structural chromosome rearrangements. *Am J Med Genet A, 132A*(1), 36-40. doi: 10.1002/ajmg.a.30378.

Kolialexi, A., Tsangaris, G. T., Kitsiou, S., Kanavakis, E., & Mavrou, A. (2005). Impact of cytogenetic and molecular cytogenetic studies on hematologic malignancies. *Anticancer Res, 25*(4), 2979-2983.

Kruzova L., S. P., Holzerova M., Vatolikova M., Divoka M., Turcsanyi P., Urbanova R., Kudelka M., Radvansky M., Kriegova E., Papajik T., Urbankova H. (2017). Gender, Age and TP53 Aberrations in Context of Complex Karyotypes in Chronic Lymphocytic Leukemia Patients in the Era of Chemo +/- Immunotherapy *Blood, 130*(Suppl 1), 2991.

Kujawski, L., Ouillette, P., Erba, H., Saddler, C., Jakubowiak, A., Kaminski, M., Malek, S. N. (2008). Genomic complexity identifies patients with aggressive chronic lymphocytic leukemia. *Blood, 112*(5), 1993-2003. doi: 10.1182/blood-2007-07-099432.

Langer-Safer, P. R., Levine, M., & Ward, D. C. (1982). Immunological method for mapping genes on Drosophila polytene chromosomes. *Proc Natl Acad Sci US, 79*(14), 4381-4385.

Levin, M., Le Coniat, M., Bernheim, A., & Berger, R. (1986). Complex chromosomal abnormalities in acute nonlymphocytic leukemia. *Cancer Genet Cytogenet, 22*(2), 113-119.

Mayr, C., Speicher, M. R., Kofler, D. M., Buhmann, R., Strehl, J., Busch, R., Wendtner, C. M. (2006). Chromosomal translocations are associated with poor prognosis in chronic lymphocytic leukemia. *Blood, 107*(2), 742-751. doi: 10.1182/blood-2005-05-2093.

McGowan-Jordan J., Simons A., & Schmid M., e. (Eds.). (2016). *An International System for Human Cytogenomic Nomenclature (2016).*

Mitelman, F. (2017). *Mitelman Database of Chromosome Aberrations and Gene Fusions in Cancer. Mitelman Database of Chromosome Aberrations and Gene Fusions in Cancer* https://omictools.com/mitelman-database-of-chromosome-aberrations-and-gene-fusions-in-cancer-tool.

Negrini, S., Gorgoulis, V. G., & Halazonetis, T. D. (2010). Genomic instability--an evolving hallmark of cancer. *Nat Rev Mol Cell Biol, 11*(3), 220-228. doi: 10.1038/nrm2858.

Nemec, P., Zemanova, Z., Kuglik, P., Michalova, K., Tajtlova, J., Kaisarova, P., ... Czech Myeloma, G. (2012). Complex karyotype and translocation t(4;14) define patients with high-risk newly diagnosed multiple myeloma: results of CMG2002 trial. *Leuk Lymphoma, 53*(5), 920-927. doi: 10.3109/10428194.2011.634042.

Olney, H. J., & Le Beau, M. M. (2001). The cytogenetics of myelodysplastic syndromes. *Best Pract Res Clin Haematol, 14*(3), 479-495. doi: 10.1053/beha.2001.0151.

Palumbo, E., & Russo, A. (2016). Chromosome Imbalances in Cancer: Molecular Cytogenetics Meets Genomics. *Cytogenet Genome Res, 150*(3-4), 176-184. doi: 10.1159/000455804.

Peterson, J. F. (2017). The Complexities of Defining a Complex Karyotype in Hematological Malignancies: A Need for Standardization? *Acta Haematol, 138*(1), 65-66. doi: 10.1159/000477894.

Pinkel, D., Segraves, R., Sudar, D., Clark, S., Poole, I., Kowbel, D., Albertson, D. G. (1998). High resolution analysis of DNA copy number variation using comparative genomic hybridization to microarrays. *Nat Genet, 20*(2), 207-211. doi: 10.1038/2524.

Pinkel, D., Straume, T., & Gray, J. W. (1986). Cytogenetic analysis using quantitative, high-sensitivity, fluorescence hybridization. *Proc Natl Acad Sci US, 83*(9), 2934-2938.

Puiggros, A., Collado, R., Calasanz, M. J., Ortega, M., Ruiz-Xiville, N., Rivas-Delgado, A., Espinet, B. (2017). Patients with chronic lymphocytic leukemia and complex karyotype show an adverse outcome even in absence of TP53/ATM FISH deletions. *Oncotarget, 8*(33), 54297-54303. doi: 10.18632/oncotarget.17350.

Rigolin, G. M., Formigaro, L., Cavallari, M., Quaglia, F. M., Lista, E., Urso, A., Cuneo, A. (2017). An extensive molecular cytogenetic characterization in high-risk chronic lymphocytic leukemia identifies karyotype aberrations and TP53 disruption as predictors of outcome and chemorefractoriness. *Oncotarget, 8*(17), 28008-28020. doi: 10.18632/oncotarget.15883.

Sarkozy, C., Terre, C., Jardin, F., Radford, I., Roche-Lestienne, C., Penther, D., Castaigne, S. (2014). Complex karyotype in mantle cell lymphoma is a strong prognostic factor for the time to treatment and overall survival, independent of the MCL international prognostic index. *Genes Chromosomes Cancer, 53*(1), 106-116. doi: 10.1002/gcc.22123.

Sawyer, J. R. (2011). The prognostic significance of cytogenetics and molecular profiling in multiple myeloma. *Cancer Genet, 204*(1), 3-12. doi: 10.1016/j.cancergencyto.2010.11.002.

Schanz, J., Tuchler, H., Sole, F., Mallo, M., Luno, E., Cervera, J., Haase, D. (2012). New comprehensive cytogenetic scoring system for primary myelodysplastic syndromes (MDS) and oligoblastic acute myeloid leukemia after MDS derived from an international database merge. *J Clin Oncol, 30*(8), 820-829. doi: 10.1200/JCO.2011.35.6394.

Schrock, E., du Manoir, S., Veldman, T., Schoell, B., Wienberg, J., Ferguson-Smith, M. A., Ried, T. (1996). Multicolor spectral karyotyping of human chromosomes. *Science, 273*(5274), 494-497.

Sikkema-Raddatz, B., Johansson, L. F., de Boer, E. N., Almomani, R., Boven, L. G., van den Berg, M. P., ... Sinke, R. J. (2013). Targeted next-generation sequencing can replace Sanger sequencing in clinical diagnostics. *Hum Mutat, 34*(7), 1035-1042. doi: 10.1002/humu.22332.

Solinas-Toldo, S., Lampel, S., Stilgenbauer, S., Nickolenko, J., Benner, A., Dohner, H., Lichter, P. (1997). Matrix-based comparative genomic hybridization: biochips to screen for genomic imbalances. *Genes Chromosomes Cancer, 20*(4), 399-407.

Speicher, M. R., Gwyn Ballard, S., & Ward, D. C. (1996). Karyotyping human chromosomes by combinatorial multi-fluor FISH. *Nat Genet, 12*(4), 368-375. doi: 10.1038/ng0496-368.

Stephens, P. J., Greenman, C. D., Fu, B., Yang, F., Bignell, G. R., Mudie, L. J., Campbell, P. J. (2011). Massive genomic rearrangement acquired in a single catastrophic event during cancer development. *Cell, 144*(1), 27-40. doi: 10.1016/j.cell.2010.11.055

Stolzel, F., Mohr, B., Kramer, M., Oelschlagel, U., Bochtler, T., Berdel, W. E., Ehninger, G. (2016). Karyotype complexity and prognosis in acute myeloid leukemia. *Blood Cancer J, 6*, e386. doi: 10.1038/bcj.2015.114.

Swerdlow S. H., Campo E., Harris N. L., et al., (Eds.). (2008). *WHO Classification of Tumours of Haematopoietic and Lymphoid Tissues.*

Thompson, P. A., O'Brien, S. M., Wierda, W. G., Ferrajoli, A., Stingo, F., Smith, S. C., Keating, M. J. (2015). Complex karyotype is a stronger predictor than del(17p) for an inferior outcome in relapsed or refractory chronic lymphocytic leukemia patients treated with ibrutinib-based regimens. *Cancer, 121*(20), 3612-3621. doi: 10.1002/cncr.29566.

Van Den Neste, E., Robin, V., Francart, J., Hagemeijer, A., Stul, M., Vandenberghe, P., Michaux, L. (2007). Chromosomal translocations independently predict treatment failure, treatment-free survival and overall survival in B-cell chronic lymphocytic leukemia patients treated with cladribine. *Leukemia, 21*(8), 1715-1722. doi: 10.1038/sj.leu.2404764.

Vardiman, J. W. (2003). Myelodysplastic syndromes, chronic myeloproliferative diseases, and myelodysplastic/myeloproliferative diseases. *Semin Diagn Pathol, 20*(3), 154-179.

Vermeesch, J. R., Fiegler, H., de Leeuw, N., Szuhai, K., Schoumans, J., Ciccone, R., Zuffardi, O. (2007). Guidelines for molecular karyotyping in constitutional genetic diagnosis. *Eur J Hum Genet, 15*(11), 1105-1114. doi: 10.1038/sj.ejhg.5201896.

Vermeesch, J. R., Melotte, C., Froyen, G., Van Vooren, S., Dutta, B., Maas, N., Devriendt, K. (2005). Molecular karyotyping: array CGH quality criteria for constitutional genetic diagnosis. *J Histochem Cytochem, 53*(3), 413-422. doi: 10.1369/jhc.4A6436.2005.

Yoshizato, T., Nannya, Y., Atsuta, Y., Shiozawa, Y., Iijima-Yamashita, Y., Yoshida, K., Ogawa, S. (2017). Genetic abnormalities in myelodysplasia and secondary acute myeloid leukemia: impact on outcome of stem cell transplantation. *Blood, 129*(17), 2347-2358. doi: 10.1182/blood-2016-12-754796.

In: An Essential Guide to Cytogenetics
Editors: N. Norris and C. Miller
ISBN: 978-1-53613-370-7


*Chapter 4*

# NATURAL HYBRIDIZATION BETWEEN CHROMOSOMAL DISCREPANT SPECIES AND THE ROLE OF HYBRID SPECIATION IN THE GENUS *ASTYANAX*

***Francisco de Menezes Cavalcante Sassi*[1]*,***
***Snaydia Viegas Resende*[1,2]*, Rubens Pazza*[1]**
***and Karine Frehner Kavalco*[1,2]**

[1]Laboratório de Genética Ecológica e Evolutiva, Universidade Federal de Viçosa, Campus Rio Paranaíba. Rio Paranaíba – MG, Brasil
[2]Programa de Pós-Graduação em Manejo e Conservação de Ecossistemas Naturais e Agrários (MCENA), Universidade Federal de Viçosa, Campus Florestal– MG, Brasil

## ABSTRACT

Initially identified as *Astyanax schubarti* on the basis of morphological characteristics, its chromosomal analysis revealed a unique diploid number in the genus. With 42 chromosomes and the impossibility of homologous pairing, the karyotype of the individual was compared to a set of haploid complements of *Astyanax schubarti* (2n = 36 chromosomes) and *Astyanax fasciatus* (2n = 48 chromosomes). Natural hybrids are rare. The viability of a hybrid between species with such chromosomal discrepancy may offer important hypotheses to explain the morphological, molecular, and cytogenetic diversity of the genus.

**Keywords**: hybridization, speciation, ecological genetics, evolutionary genetics

## INTRODUCTION

Natural hybrids emerge following mating between individuals of two populations or distinct species, and they are identified on the basis of one or more hereditary characteristics (Downling and Secor 1997). The role of hybridization in the evolution of species remains controversial, but hybrid speciation, mainly in polyploid plants, is well documented (Mallet 2007). This phenomenon is common among species of teleost fish (Scribner et al., 2001; Salzburger et al., 2002; Meyer et al., 2005), but few studies have focused on the species of the Neotropical region, particularly in South America.

*Piabas* or *lambaris* in Brazil are representatives of fish belonging to the genus *Astyanax*, which is most abundant in South America (Géry 1977). Among the about 150 species already described, *A. mexicanus* stands out, and because it has cave populations with troglomorphic characteristics, it is used as a model species in developmental and evolutionary biology studies (Jeffery 2001). In South America, where most of the biodiversity of this genus occurs, chromosomal studies have identified several cryptic species in the last 25 years, including the groups *A. scabripinnis* (2n = 46, 48, and 50), *A. fasciatus* (2n = 46 and 48), and *A. bimaculatus* (2n = 50). In addition to the differences in diploid numbers, these groups also present variation in karyotype forms and in the presence of supernumerary chromosomes (Pazza and Kavalco 2007).

Most species of *Astyanax* with a characterized chromosome number have 46 to 50 chromosomes, except for *A. schubarti*, which has 2n = 36 chromosomes, possibly resulting from Robertsonian translocations from an ancestral karyotype (Daniel-Silva and Almeida-Toledo 2005). *A. fasciatus* can present 2n = 46 or 48 chromosomes in sympatric or allopatric populations (Artoni et al., 2006; Pazza et al., 2006; Pazza et al., 2008; Pansonato-Alves et al., 2013; Penteado et al., 2013; Pazza et al., 2016). In the Mogi-Guaçu River (upper Rio Paraná Basin), the largest variation in sympatric chromosomes is observed in the species, where in addition to the standard cytotypes (with adequate homology between the chromosomes) 2n = 46 and 48, specimens with 2n = 45, 46 variants, and 47 chromosomes have also been observed, without supernumerary chromosomes (Pazza et al., 2006). These authors suggest that the chromosome number 2n = 48 is common in the Mogi-Guaçu River basin, whereas the chromosome number 2n = 46 represents an invasion from the Tietê River basin, and that secondary contact after the recent diversification enabled gene flow between the cytotypes, despite the non-identification of obvious F1 hybrids in the study.

Thus, the present study, which presents a hybrid between species with divergent karyotypes, *A. schubarti* and *A. fasciatus*, provides important data for future studies on the natural history of the group.

## MATERIAL AND METHODS

Samples of the genus *Astyanax* were collected in the Mogi-Guaçu River (143 individuals), in the region of Cachoeira de Emas (Pirassununga, São Paulo) between 2001 and 2003 (-21.92695, -47.3673). Animals were fixed in formalin 10%, put in in ethanol 70% and deposited in the vertebrate collection of the Laboratory of Ecological and Evolutionary Genetics of the Federal University of Viçosa, Rio Paranaíba Campus (LaGEEvo UFV/CRP) - MG, Brazil. The tissue samples used for molecular analyses were stored at -20 °C in pure ethanol. Species identification follow the diagnostic characters for this species, like teeth morphology, fin color after fixation and meristic.

Mitotic chromosomes were obtained by the air-drying technique (Bertollo et al., 1978) from the anterior kidney of the animal. Chromosomes were analyzed after conventional staining with 1% Giemsa, and at least 20 metaphases were counted. The chromosomes were classified as metacentric (m), submetacentric (sm), subtelocentric (st), and acrocentric (a), according to the method described by Levan et al., (1964), using Photoshop CC v3.5 for measurement, and were later organized into karyotypes, according to the convention for fish.

Total DNA was extracted from each individual using a commercially available kit and fragments of organs, such as liver and heart, deposited in the Tissue Bank of the Laboratory of the Ecological and Evolutionary Genetics (LaGEEvo), Federal University of Viçosa, Rio Paranaíba *Campus*.

The mitochondrial gene (mtDNA) cytochrome oxidase I (COI) was amplified with the primers Fish R1 - 5′ TAGACTTCTGGGTGGCC AAAGAATCA3′ and Fish F1 - 5′ TCAACCAACC ACAAAGACATTGG CAC3′ (Ward et al., 2005). PCR was performed in a final volume of 25 μL, with 2.5 μL of 10X Taq buffer, 1 μL $MgCl_2$, 1 μL of each primer, 0. 2μL of Taq DNA polymerase, 12.8 μL ultra-pure water, 1.5 μL of dNTPs, and 5 μL of the extracted DNA solution. Amplifications were performed in a thermal cycler as follows: initial denaturation at 95 °C for 2 min; and repeated cycles of denaturation at 94 °C for 30 s, primer annealing st 58 °C for 30 s, and a final extension at 72 °C for 1 min. PCR products were visualized on a 1% agarose gel and samples were sent to a third-party company (Macrogen Korea) for purification and sequencing.

Once the sequences were obtained, they were identified using the BLASTn algorithm in the GenBank database (http://blast.ncbi.nlm. nih.gov), which is used to compare nucleotide sequences.

## RESULTS

One individual (Figure 1) identified as *A. schubarti* Britski (1964), presented 42 chromosomes (10 m + 19 sm + 8 st + 5 a) (Figure 2). It was not possible to completely pair the chromosomes, since measurements showed that the size of each chromosome is different, and morphology too, indicating that there is no homology.

Figure 1. Photograph of the hybrid between *Astyanax fasciatus* and *Astyanax schubarti* fixed in 10% formalin and preserved in 70% alcohol. Standard length = 80.89mm.

Figure 2. Karyotype of the hybrid stained with Giemsa.

Analysis of genetic identity using BLAST indicated 99% similarity in the gene encoding *cytochrome oxidase I* from the study specimen with most of the sequences of the same gene from *A. fasciatus* present in the database.

## DISCUSSION

In this study, we report the first natural hybrid of the genus *Astyanax*. Unlike other natural hybrids found in fish, it is necessary to emphasize discrepancies in the diploid number of parent species. The contrary hypothesis would be a completely new cytotype for the genus *Astyanax* in more than 40 years of cytogenetics studies in the genus, but it is not plausible since we have found only one specimen among more than 240 individuals analyzed.

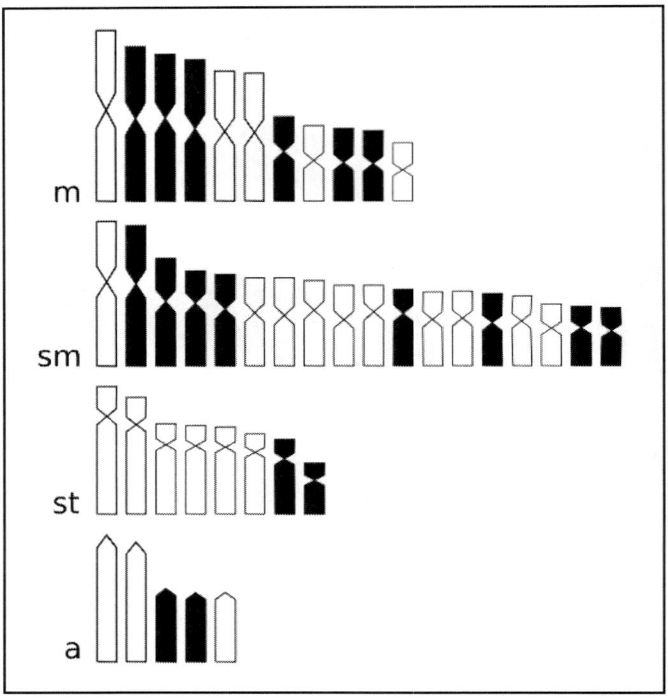


Figure 3. Ideogram representing the probable parental origin of the hybrid chromosomes between *Astyanax schubarti* (in black) and *A. fasciatus* (in white).

As the chromosome number was not consistent with that of any other *Astyanax* species, we tested the hypothesis that it represented a hybrid of *A. schubarti* (2n = 36, 12 m + 16 sm + 4 st + 4 a) and *A. fasciatus* (2n = 48, 8 m + 22 sm + 12 st + 6 a), which are both sympatric and syntopic species in

Cachoeira de Emas. Hence, an ideogram with the haploid chromosome complement of each species was assembled from karyotype data in the literature (Daniel-Silva & Almeida-Toledo 2005; Pazza et al., 2006), and the result (Figure 3) was congruent with the karyotype obtained.

Other natural, or even artificial hybrids, of Neotropical species have parents with at least the same chromosome number. Parents of the natural hybrid between species of the genus *Cichla* (*Cichla monoculus* and *C. temensis*) feature 2n = 48 chromosomes (Brinn et al., 2004), with little difference in the location of the nucleolus organizer region. Natural hybrids of *Pseudoplatystoma* species have already been observed in some populations using molecular markers (Prado et al., 2012a; Carvalho et al., 2013). However, no cytogenetic differences have been observed between *Pseudoplatystoma corruscans* and *P. reticulatum*, which both have 2n = 56 chromosomes, between parents and the artificial hybrid (Prado et al., 2012b). The same can be observed in the artificial hybrid between *Piaractus mesopotamicus* and *Colossoma macropomum,* both with 2n = 54 chromosomes, and one of the oldest produced in Brazil (Almeida-Toledo et al., 1987). At that time, both species were classified under the same genus. In turn, the hybrid between *Colossoma macropomum* and *Piaractus brachypomus* (also with 2n = 54 chromosomes in both species) can be recognized by the location of the 18S ribosomal gene, because it has four loci distributed in its chromosomal complement while the parents have two and six, respectively (Nirchio et al., 2003). In the *Astyanax* genus, artificial hybrids between surface and cave populations of *A. mexicanus* have been developed for studies investigating the development biology of the eye (Wilkens 1971). Furthermore, mitochondrial DNA sequences suggest introgression events between these forms, which are currently geographically isolated (Dowling et al., 2002). Although no chromosomal data are available, it is likely that both populations have a similar chromosomal complement (Kavalco and Almeida-Toledo 2007), with 2n = 50 chromosomes. Thus, an intermediate diploid hybrid between the parent species *A. fasciatus* (2n = 48) and *A. schubarti* (2n = 36) with significantly different diploid numbers is remarkable.

Although being characterized as common in fish (Scribner et al., 2001), natural hybrids in Neotropical fish are scarcely reported. Among the 168 species reported by those authors, none is from South America, although the literature indicates a hybrid between *P. mesopotamicus* and *C. Macropomum* as described previously (Almeida-Toledo et al., 1987). The difficulty in identification of hybrids in nature can be explained by the sterility of interspecific hybrids (Anderson 1948). However, difficulties in identification can be related to morphological characteristics, since these are used initially to identify species. Although some variables are intermediate in hybrids, their distribution is not always uniform, according to multivariate analysis (Neff and Smith 1979), leading the hybrid to present more characteristics of one of the parents. In fact, the hybrid found in the present study was initially identified as *A. schubarti*, due to the presentation of classic morphological characteristics, differing from *A. fasciatus* by the height of the body and the absence of red pigmentation of the caudal and dorsal fins (Britski 1964), and from *A. altiparanae* by the absence of an oval humeral spot (Garutti and Britski 2000). This suggests that the low frequency of hybrids documented in the Neotropical region can be related to the subtlety of morphological characteristics presented by hybrids, impeding their observation when only morphological data are considered.

One of the bets of current science against problems such as this is DNA barcoding, which involves the analysis of a specific mitochondrial sequence (COI) for the taxonomic identification of unknown individuals (Hebert et al., 2003). Analysis of this sequence in hybrid individuals enables the identification of *A. fasciatus*, which suggests that this is the female parental species by mitochondrial inheritance. However, it is important to highlight that DNA barcoding may not always be effective, especially for cryptic species, often with hybridism or with small sample sizes (Moritz and Cicero 2004; Meyer and Paulay 2005). The latter seems to be resolved in the genus *Astyanax*, which has about 150 described species and for which DNA barcoding has identified up to five key groups and several potential new species (Rossini et al., 2016). According to these authors, *A. fasciatus* and *A. schubarti* belong to clade 1 and display the lowest interspecific distance, with clusters including more than one nominal species. The low divergence observed in this clade may be explained by phenotypic plasticity, which hinders the identification of species; different rates of evolution in the *COI* gene between groups, with different groups having more recent adaptive radiation than others; and the description of new species only by their restricted geographical distribution, which would require future studies for synonymization (Rossini et al., 2016). Thus, despite the possibility of reducing gaps with larger sampling, resolution of the molecular taxonomy of *Astyanax* by DNA barcoding can be compromised by species complexes and hybridism.

The viability of a hybrid between species with such disparate chromosome numbers can help in our understanding of hybridity as an important factor for diversification in *Astyanax*. The fact that the individual morphologically of the hybrid resembles one of the parents may aid in the introgression of the genome in the population, since some species of fish may recognize their partners by morphology, as in the case of African cichlids (Kocher 2004). The speed and direction of evolution in new environments can be directly influenced by introgression, as this increases genetic variation (Lu et al., 1995; Abbott et al., 2003) and can produce new genotypes (Svardson 1970; Rieseberg et al., 2003). This may facilitate evolutionary divergence and speciation by reorienting the evolution of populations when there are changes in the environment (Grant et al., 1995). Although the emergence of a natural hybrid is rare, this is not related to evidence of the insignificance of hybridization and introgression in the evolution and diversification of animals (Downling & Secor 1997), since even small rates of hybridization can have a great impact on the evolution of a species (Mallet 2005). A recent adaptive radiation (Ornelas-Garcia et al., 2008) involving morphological and chromosomal variation, can therefore explain the viability in hybrids and their fundamental role in the evolution of the *Astyanax* genus. In addition, this also explains the introgressions and the shared haplotypes between different species observed by molecular data.

Additionally, Scribner et al., (2001) noted four main factors for hybridization in fish: habitat loss, distribution expansion, aquaculture, and introduction. These factors can be easily attributed to anthropogenic action. In view of this, more and more human endeavors aimed at altering aquatic habitats must be monitored using multidisciplinary approaches, including molecular and cytogenetic markers, so that the management of possible anthropological hybridization events is minimized. We conclude that hybridisms between species with discrepant chromosomal numbers can occur in nature and could have great importance in *Astyanax* evolution. More studies are necessaries to understand the impact of such events in evolution of this group.

## Acknowledgments

The authors thank to FAPEMIG (Fundação de Amparo à Ciência do Estado de Minas Gerais) and to Post-graduation Program MCENA/UFV (Manejo e Conservação de Ecossistemas Naturais e Agrários).

## References

Abbot, R. J., James, J. K., Milne, R. I., Gillies, A. C. M. (2003) Plant introductions, hybridization and gene flow. *Philos. Trans. R. Soc. Lond. B., Biol. Sci.* 358: 1123-1132.

Almeida-Toledo, L. F., Foresti, F., Toledo Filho, S. A., Bernardino, G., Ferrari, W., Alcantara, R. C. G. (1987) Cytogenetic studies of Colossoma mitrei, Colossoma macropomum and their interspecific hybrid. *Proc. World Symp. Selection Hybrid Genet. Eng. Aquac.* 189-195.

Anderson, E. (1948) Hybridization of the habitat. *Inter. Journ. Organ.* Evol. 1: 1-9.

Artoni, R. F., Shibatta, O. A., Gross, M. C., Schneider, C. H., Almeida, M. C., Vicari, M. R., Bertollo, L. A. C. (2006) Astyanax aff. Fasciatus Cuvier, 1819 (Teleostei; Characidae): evidences of a species complex in the Upper rio Tibagi basin (Paraná, Brazil). *Neotrop. Ichtyol.* 4: 197-202.

Bertollo, L. A. C., Takahashi, C. S., Moreira-Filho, O. (1978) Cytotaxonomic considerations on Hoplias lacerdae (Pisces, Erythrinidae). *Braz. J. Genet.* 1: 103-120.

Brinn, M. N. A., Porto, J. I. R., Feldberg, E. (2004) Karyological evidence for interspecific hybridization between Cichla monoculus and C. temensis (Perciformes, Cichlidae) in the Amazon. *Hereditas.* 141: 252-257.

Britski, H. A. (1964) On a new species of *Astyanax* of the river Mogi-Guassu (Pisces, Characidae). *Pap. Avuls. Depart. Zool.* 16: 213-215.

Carvalho, D. C., Seerig, A. S., Brasil, B. S. A. F., Crepaldi, D. V., Oliveria DAA (2013) Molecular identification of the hybrid between the catfish species Pseudoplatystoma corruscans and Pseudoplatystoma reticulatum using a set of eight microsatellite markers. *J. Fish. Biol.* 83: 671-676.

Daniel-Silva, M. F. Z., Almeida-Toledo, L. F. (2005) Chromosome evolution in fish: BrdU replication patterns demonstrate chromosome homeologies in two species of the genus *Astyanax*. *Cytog. Gen. Res.* 109: 497-501.

Dowling, T. E. and Secor, C. L. (1997) The role of hybridization and introgression in the diversification of animals. *Ann. Rev. Eco. System.* 28: 593-619.

Dowling, T. E., Martasian, D. P., Jeffery, W. R. (2002) Evidence for multiple genetic forms with similar eyeless phenotypes in the blind cavefish, *Astyanax mexicanus*. *Mol. Biol. Evol.* 19: 446-455.

Eigenmann, C. H. (1921). The American Characidae. *Mem. Museum Comp. Zool.* 43: 209–310.

Garutti, V. and Britski, H. A. (2000) Description of a new species of *Astyanax* (Teleostei: Characidae) from the upper Paraná River basin and considerations about the other species of the genus in the basin. *Comum. Museu Cienc. Tecnol.* 13: 65–88.

Géry, J. (1977) *Characoids of the world.* 1st ed. TFH Publications Inc: Neptune.

Grant, P. R., Grant, B. R., Petren, K. (1995) Hybridization in the recent past. *Am. Nat.* 166: 56-67.

Hebert, P. D. N., Cywinska, A., Ball, S. L., deWaard, J. R. (2003) Biological identifications through DNA barcodes. *Proc. Royal. Soc. B.* 270: 313-321.

Jeffery, W. R. (2001) Cavefish as a model system in evolutionary developmental biology. *Dev. Biol.* 2001, 231: 1-12.

Kavalco, K. F., Almeida-Toledo, L. F. (2007) Molecular cytogenetics of blind Mexican tetra and comments on the karyotypic characteristics of genus *Astyanax* (Teleostei, Characidae). *Zebrafish*. 4: 103-111.

Kavalco, K. F., Pazza, R., Brandão, K. O., Garcia, C., Bertollo, L. A. C., Almeida-Toledo, L. F. (2016) Chromosomal diversification higher than molecular variation in *Astyanax aff. Fasciatus* (Teleostei, Characidae). *Zebrafish*. 13: 345-353.

Kocher, T. D. (2004) Adaptive evolution and explosive speciation: the cichlid fish model. *Nature Rev. Genet.* 5: 288-298.

Levan, A., Fredga, K., Sandberg, A. A. (1964) Nomenclature of centromeric position on chromosomes. *Hereditas*. 52: 201-220.

Lu, M., Szmidt, A. E., Wang, X. (1995) Inheritance of RAPD fragments in haploid and diploid tissues of *Pinus sylvestris* (L.). *Heredity*. 74: 582-589.

Mallet, J. (2005) Hybridization as an invasion of the genome. *Tren. Ecol. Evol.* 20: 229-237.

Mallet, J. (2007) Hybrid speciation. *Nature.* 446: 279-283.

Meyer, C. P., Paulay, G. (2005) DNA Barcoding: error rates based on comprehensive sampling. *Plos. Biol*. 3: 2229-2238.

Moreira, Filho, O., Bertollo, L. A. C. (1991) Astyanax scabripinnis (Pisces, Characidae): a species complex. *Brazilian J. Genet.* 14: 331-357.

Moritz, C., Cicero, C. (2004) DNA Barcoding: Promise and Pitfalls. *Plos. Biol.* 2: e354.

Neff, N. A., Smith, G. R. (1979) Multivariate analysis of hybrid fishes. *System Biol.* 28: 176-196.

Nirchio, M., Fenocchio, A. S., Swarça, A. C., Pérez, J. E., Granado, A., Estrada, A., Ron, E. (2003) Cytogenetic characterization of hybrids offspring between *Colossoma macropomum* (Cuvier, 1818) and *Piaractus brachypomus* (Cuvier, 1817) from Caicara del Orinioco, Venezuela. *Caryologia.* 56: 405-411.

Ornelas-Garcia, C. P., Domínguez-Domínguez, O., Doadrio, I. (2008) Evolutionary history of the fish genus Astyanax Baird & Girard (1854) (Actinopterygii, Characidae) in Mesoamerica reveals multiple morphological homoplasies. *BMC Evol. Biol.* 8: 340.

Pansonato-Alves, J. C., Hilsdorf, A. W. S., Utsunomia, R., Silva, D. M. Z. A., Oliveira, C., Foresti, F. (2013) Chromosomal mapping of repetitive DNA and Cytochrome c Oxidase I sequence analysis reveal differentiation among sympatric samples of *Astyanax fasciatus* (Characiforems, Characidae). *Cytogenet. Genome Res.* 141: 133-142.

Pazza, R., Argento, R. F. M., Mariano, I. H. S., Kavalco, K. F. (2016) Phenotypic plasticity and karyotype conservation in allopatric populations of *Astyanax aff. Fasciatus* (Teleostei, Characiformes) at the Upper Paraná River basin. *Acta Zool.* 97: 241-245.

Pazza, R., Kavalco, K. F. (2007) Chromosomal evolution in the Neotropical characin *Astyanax* (Teleostei, Characidae). *Nucleus.* 50: 519-543.

Pazza, R., Kavalco, K. F., Bertollo, L. A. C. (2006) Chromosome polymorphism in Astyanax fasciatus (Teleostei, Characidae). 1. Karyotype analysis, Ag-NORs and mapping of the 18S and 5S ribosomal genes in sympatric karyotypes and their possible hybrid forms. *Cytogenet. Genome Res.* 112: 313-319.

Pazza, R., Kavalco, K. F., Prioli, S. M. A. P., Prioli, A. J., Bertollo, L. A. C. (2007) Chromosome polymorphism in Astyanax fasciatus (Teleostei, Characidae), Part 3: Analysis of the RAPD and ISSR Molecular markers. *Biochem. Syst. Ecol.* 35: 843-851.

Pazza, R., Kavalco, S. A. F., Penteado, P. R., Kavalco, K. F., Almeida-Toledo, L. F. (2008) The species complex *Astyanax fasciatus Cuvier* (Teleostei, Characiformes) – a multidisciplinary approach. *J, Fish. Biol.* 72: 2002-2010.

Penteado, P. R., Kavalco, K. F., Pazza, R. (2013) A third cytotype of *Astyanax aff. Fasciatus* (Teleostei, Characidae) from the São Francisco river basin. *Evol. Cons. Biodivers.* 4: 1-7.

Prado, F. D., Hashimoto, D. T., Senhorini, J. A., Foresti, F., Porto-Foresti, F. (2012a) Detection of hybrids and genetic introgression in wild stocks of two catfish species (Siluriformes: Pimelodidae): The impact of hatcheries in Brazil. *Fisher Res.* 125-126: 300-305.

Prado, F. D., Nunes, T. L., Senhorini, J. A., Bortoluzzi, J., Foresti, F., Porto-Foresti, F. (2012b) Cytogenetic characterization of F1, F2 and backcross hybrids of the Neotropical catfish species Pseudoplatystoma corruscans and Pseudoplatystoma reticulatum (Pimelodidae, Siluriformes). *Genetic Mol. Biol.* 35: 57-64.

Rieseberg, L. H., Raymond, O., Rosenthal, D. M., Lai, Z., Livingstone, K., Nakazato, T., Durphy, J. L., Schwarzbach, A. E., Donovan, L. A., Lexer, C. (2003) Major ecological transitions in wild sunflowers facilitated by hybridization. *Science.* 29: 1211-1216.

Rossini, B. C., Oliveira, C. A. M., Melo, F. A. G., Bertaco, V. A., Astarloa, R. N. D., Rosso, J. J. et al., (2016) Highlighting Astyanax species diversity through DNA Barcoding. *Plos. One.* 11: e0167203.

Salzburger, W., Baric, S., Sturmbauer, C (2002) Speciation via introgressive hybridization in East African cichlids. *Mol. Ecol.* 11: 619-625.

Santos, V. H., Foresti, F., Oliveira, C., Almeida-Toledo, L. F., Toledo-Filho, S. A., Bernardino, G. (2002) Synaptonemal complex analysis in the fish species Piaractus mesopotamicus and Colossoma macropomum, and in their interspecific hybrid. *Caryologia.* 55: 73-79.

Scribner, K. T., Page, K. S., Bartron, M. L. (2001) Hybridization in freshwater fishes: a review of case studies and cytonuclear methods of biological inference. *Rev. Fish. Biol. Fisher.* 10: 293-323.

Svärdson, G. (1970) Significance of introgression in coregonid evolution. *Biology of Coregonid fishes*. 33-59.

Ward, R. D., Zemlak, T. S., Innes, B. H., Last, P. R., Hebert, P. D. N. (2005) DNA barcoding Australia's fish species. *Philos Trans Royal Soc*. B. 360: 1847–1857.

Wilkens, H. (1971) Genetic interpretation of regressive evolutionary processes: studies on hybrid eyes of two *Astyanax* cave populations (Characidae, Pisces). *Evolution*. 25: 530-544.

In: An Essential Guide to Cytogenetics ISBN: 978-1-53613-370-7
Editors: N. Norris and C. Miller 

*Chapter 5*

# TRICHORHINOPHALANGEAL SYNDROME (TRPS) REVISITED

***N. Selenti**, *M. Tzetis, E. Tsoutsou and H. Fryssira***

Department of Medical Genetics,
National and Kapodistrian University of Athens,
School of Medicine, Aghia Sophia Children's Hospital,
Athens, Greece

## ABSTRACT

There are three distinct subtypes of Trichorhinophalangeal syndrome (TRPS); TRPS type I, TRPS type II and TRPS type III. Features common to all three subtypes include sparse, slowly growing scalp hair, laterally sparse eyebrows, a bulbous tip of the nose (pear-shaped), and protruding ears. The diagnosis of TRPS is based in typical clinical and radiographic features, as well as in the identification of causative mutations in the *TRPS1* gene (TRPS type I and III) and in loss of functional copies of the *TRPS1* and EXT1 genes (TRPS type II). The mode of inheritance in TRPS I and III is autosomal dominant, however *de novo* deletions of *TRPS1* and EXT1 genes are the main defect of TRPS II. Parental

* Corresponding Author Email: nikoletta_selenti@hotmail.com.

balanced chromosomal rearrangements are an important cause of interstitial aberrations in TRPS. In cases of cytogenetically-invisible alterations, parental FISH analysis as well as aCGH should be considered as part of the clinical baseline testing. Treatment of clinical problems of TRPS types is mainly supportive and includes ectodermal and skeletal issues. The number of distinct syndromes as TRPS is rapidly increasing and their confirmation is necessary especially in cases that typical features are absent. Clinical geneticists should provide information for the families and advise them how to overcome problems.

**Keywords**: trichorhinophalangeal syndrome (TRPS), genotype/phenotype of TRPS patients, differential diagnosis of TRPS

## INTRODUCTION

Trichorhinophalangeal Syndrome (TRPS) is a rare inherited multisystem disorder. This is an autosomal dominant malformation syndrome, which is characterized by craniofacial and skeletal abnormalities, although more than half of the reported cases are sporadic. TRPS exhibits almost complete penetrance, but variable expressivity. Phenotype can vary depending on age and gender, even among patients carrying identical mutations, affected members of the same family, or in monozygotic twins. There are three distinct subtypes TRPS I, II and III and their diagnosis is based on clinical and radiographic features as well as on genetic analysis that is helpful especially in the case of non-classical clinical presentation [1-3].

### Phenotype of TRPS Type I, II and III Patients

Clinical features that are common to all subtypes include sparse scalp hair, lateral eyebrows, bulbous tip of the nose, long flat philtrum, thin upper vermilion border, and protruding ears [1, 3].

### *TRPS Type I Patients*

TRPS type I is caused by mutations in the *TRPS1* gene located on chromosome 8q24.1 and characterized by distinctive skeletal abnormalities and craniofacial dysmorphism. The name of the condition describes some of the areas of the body that are commonly affected: hair (tricho-), nose (rhino-), and fingers and toes (phalangeal) [3, 4].

TRPS I patients, besides the common clinical features, have abnormalities of the skin, teeth, sweat glands, and nails. Skeletal abnormalities include cone-shaped epiphyses at the phalanges, hip malformations and short stature. Affected individuals often have short feet with toenails and fingernails typically thin and abnormally formed. Hip dysplasia often develops in early adulthood but can occur in infancy or childhood. Children with TRPS I often present hypermobility in many of their joints, however, the joints may degenerate leading to joint pain and a limited range of joint movement [3-5].

### *TRPS Type II Patients*

TRPS II or Langer–Giedion syndrome (LGS) is considered to be an autosomal dominant condition however most cases of TRPS II are not inherited, but occur as random events -*de novo*- during the formation of reproductive cells in a parent of an affected individual [6].

The characteristic phenotype of individuals with TRPS II includes sparse scalp hair, thick eyebrows, bulbous tip of the nose, long flat philtrum, thin upper vermilion border and small teeth either with oligodontia or are supernumerary. Most individuals with TRPS II have mild intellectual disability [5, 6].

Bone and joint malformations are present too. Affected individuals may develop a few to several hundred osteochondromas that can cause pain, limited range of joint movement, or damage to the spinal cord, depending on the location of the osteochondromas. These bone growths typically begin from infancy to early childhood and stop forming around adolescence. TRPS II patients may also exhibit osteopenia as well as short stature, cone-shaped epiphyses at the phalanges, an unusually large range of joint movement (hypermobility) and hip malformations [3, 6-8].

***TRPS Type III Patients***

Trichorhinophalangeal syndrome type III (TRPS3), also known as Sugio-Kajii syndrome, is considered as an autosomal dominant inheritance and characterized by bulbous tip of the nose, long flat philtrum, thin upper vermilion border, thin light-colored hair and delayed eruption of teeth. Skeletal abnormalities include cone-shaped epiphyses at the phalanges, brachydactyly, metacarpophalangeal shortening, skeletal dysplasia and short stature. Affected individuals may also exhibit additional skeletal abnormalities including osteochondritis, thoracic scoliosis, abnormal prominence of the breast bone (pectus carinatum), and/or limited movements of certain joints. In addition, some females affected by this disorder may exhibit hip malformations. The range and severity of symptoms may vary from case to case [3, 5, 9].

## GENETICS

TRPS type I and III are caused by mutations in the *TRPS1* gene, which is located on 8q24.1. TRPS III represents only the extreme clinical manifestations of TRPSI. Missense mutations in exon 6 of *TRPS1* gene are the most common molecular defect identified in TRPSIII [2].

The *TRPS1* gene encodes a large nuclear transcription factor, protein *TRPS1*, which consists of 1281 amino acids. It binds specifically to GATA sequences and represses expression of GATA-regulated genes at selected sites and stages in vertebrate development. In its structure it combines nine potential zinc-finger motifs of four different types, including DNA-binding GATA motif and IKAROS-like zinc-finger motif, which mediates the transcriptional repressive function at the carboxy terminus. *In vitro* studies demonstrated that the presence of the DNA-binding domain is indispensable for the *TRPS1* repression function. *TRPS1* also regulates chondrocyte proliferation and differentiation and executes multiple functions in proliferating chondrocytes, expanding the region of distal chondrocytes, activating proliferation in columnar cells and supporting the differentiation of columnar into hypertrophic chondrocytes. *TRPS1* transcription factor is a novel transcriptional repressor of the *RUNX2* promoter by identifying single nucleotide variation. *RUNX2* is the master regulator of osteoblast differentiation and is required for chondrocyte hypertrophy. *TRPS1* repression activity is necessary for the timely progression of chondrocyte maturation and synchronization of chondrocyte development with perichondrial mineralization [2, 10, 11].

TRPS type I is mainly caused by entire *TRPS1* deletions and nonsense mutations located before the coding region for the DNA-binding domain. TRPS type III is caused by missense mutations in the DNA-binding domain [10].

The LGS/TRPS type II is caused by contiguous deletion of the *TRPS1* and *EXT1* (exostosin-1) genes. The multiple hereditary exostoses (MHE) type I that is present in TRPS type II is caused by a mutation in *EXT1*. The MHE is a genetically heterogeneous disorder which can be caused by mutations in the *EXT1*, *EXT2* or *EXT3* gene. EXT1 protein is a glycosyltransferase required for the biosynthesis of heparan-sulfate that is found in Golgi apparatus. It modifies newly produced enzymes and other proteins. EXT1 protein binds to EXT2 protein to form a complex, because the EXT1/EXT2 complex possesses substantially higher glycosyltransferase activity than EXT1 or EXT2 alone [12, 13].

Although deletions of the 8q region are rare and considerably variable, the shortest region of deletion overlap (SRO) has been defined at 8q24.1, spanning 2 Mb, including *TRPS1, EIF3S3, RAD21, OPG, CXIV* and *EXT1* genes [14].

Cornelia de Lange syndrome-4 (CdLs-4) that is characterized by growth deficiency, mental retardation, microcephaly, bushy eyebrows and synophrys, depressed nasal bridge, micrognathia, micromelia, hearing loss, anteverted nares, prominent symphysis and spurs in the anterior angle of mandible and gastrointestinal problems is caused by heterozygous mutations of the *RAD21* gene [15, 16].

TRPS type I, LGS/TRPS type II and CdLs-4 patients share many facial features. There are TRPSII patients where deletions in 8q24.1 region have been identified involving *RAD21* and not *TRPS1*. The phenotypic consequences of molecular defects in *RAD21*, such as facial dysmorphisms, might be underestimated in these patients [1-3, 15, 16].

# DIAGNOSIS

## Clinical Diagnosis

Facial, ectodermal features and skeletal findings are characteristics of the diagnosis of TRPS type I, LGS/TRPS type II and TRPS type III patients. For TRPS type II patients multiple osteochondromas and intellectual disability are also present in order to confirm the clinical diagnosis [1-3].

## Genetic Diagnosis

Sequence analysis of *TRPS1* is initially performed. Chromosome microarray analysis (CMA), using either aCGH or SNP arrays, or gene-targeted deletion/duplication analysis is recommended in the case of no pathogenic variant detection in *TRPS1* [2, 4, 10].

Since happloinsufficiency of *TRPS1* is responsible for the phenotype of TRPS type I and deletion of copies of the *TRPS1* and *EXT1* is responsible for the phenotype of TRPS type II, karyotype may be recommended in order to detect apparently balanced translocaton. Most chromosomal aberrations are unbalanced, resulting in duplication or deletion of genetic material at the chromosomal breakpoints. The frequency of balanced chromosomal insertions is much lower than reciprocal translocations because it requires three chromosomal breaks. Parental balanced insertions increase the possibility of copy number imbalances in the offspring resulting in congenital anomalies of multiple organ systems and psychomotor retardation. aCGH analysis revealed that approximately 2.1% of parents with offspring having developmental anomalies, harbor balanced insertions. In this regard, genomic analysis of the parental DNA of a patient may help us to exclude the possibility of parental balanced translocations as predisposing genetic risk [17, 18].

## Differential Diagnosis

TRPS is often considered in the differential diagnosis of disorders with abnormalities of the hair, nose, and limbs (Table 1) [19-21].

**Table 1. Disorders to consider in case of the differential diagnosis of the TRPS**

| Disorder | Genes | Mode of Inheritance | Clinical Features | |
|---|---|---|---|---|
| | | | Overlapping with TRPS | Distinguishing from TRPS |
| Oculo-dento-digital syndrome | *GJA1* | Autosomal Dominant | • Slow-growing, dry hair<br>• Underdeveloped alae nasi<br>• Long Philtrum | • Eye deformities<br>• Dental abnormalities |
| Cartilage-hair syndrome | *RMRP* | Autosomal Recessive | • Short stature<br>• Fine hair<br>• Cone-shaped epiphyses | • Nasal shape<br>• Immunodeficiency |
| Ellis-Van Creveld Syndrome | *EVC1*, *EVC2* | Autosomal Recessive | • Short stature<br>• Brachydactyly | • Nasal shape<br>• Oral frenula<br>• Polydactyly |

## Treatment

The treatment of TRPS is mainly supportive. Concerning the ectodermal features, practical advice on hair care and the extraction of supernumerary teeth can be considered. Human growth hormone therapy may be recommended for skeletal disorders such as short stature, however reported results vary. Regular simple analgesia is the mainstay treatment for joint paint. Physiotherapy and occupational therapy may ameliorate the mobility. Prosthetic hip implantation should be recommended in those with severe hip dysplasia. Regarding osteopenia, bisphosphonates can be given in individuals with TRPS I and bone fragility. The resection of exostoses should be considered in TRPS patients with restricted range of motion, or nerve compression. Skeletal and orthopedic symptoms continue to be an important issue. Severe scoliosis due to joint laxity, Perthes disease, asymmetry of legs, joint immobility and adjacent exostoses are among the most frequent problems. On the other hand, eye, ear, and cardiac problems are rarely of major relevance [1, 5].

## Genetic Counseling

Because of the intrafamilial clinical variability, it is not possible to predict the exact phenotype in family members who have inherited a *TRPS1* pathogenic variant or a deletion that includes *TRPS1*. However, once the genetic alteration that causes the TRPS has been identified in an affected family member, prenatal testing and/or Preimplantation Genetic Diagnosis (PGD) for TRPS is suggested. On the other hand, timely diagnosis of parental balanced chromosomal rearrangements using the appropriate techniques is important and can reduce the risk of subsequent miscarriages as well as abnormal offspring [1, 5, 18].

With better knowledge on the long-term course, complications and molecular defects that cause TRPS type I, II and III, genetic counseling is necessary by clinical geneticists [5].

## References

[1] Gorlin R. J., Cohen M. M. Jr., Wolfson J. Trichorhinophalangeal syndrome. *Am J Dis Child.* 1969, 595-599.

[2] Lüdecke H.-J., Schaper J., Meinecke P., Momeni P., Groß S., von Holtum D., Hirche H., Abramowicz M. J., Albrecht B., Apacik C.,

Christen H.-J., Claussen U., Devriendt K., Fastnacht E.,. Forderer A., Friedrich U., Goodship T. H. J., Greiwe M., Hamm H., Hennekam R. C. M., Hinkel G. K., Hoeltzenbein M., Kayserili H., Majewski F., Mathieu M., McLeod R., Midro A. T., Moog U., Nagai T., Niikawa N., Ørstavik K. H., Plöchl E., Seitz C., Schmidtke J., Tranebjærg L., Tsukahara M., Wittwer B., Zabel B., Gillessen-Kaesbach G. and Horsthemke[1] B. Genotypic and Phenotypic Spectrum in Trichorhinophalangeal Syndrome Types I and III. *Am J Hum Genet*. 2001 January, 81-91.

[3] Maas S. M., Shaw A. C., Bikker H., Lüdecke H. J., van der Tuin K., Badura-Stronka M., Belligni E., Biamino E., Bonati M. T., Carvalho D. R., Cobben J., de Man S. A., Den Hollander N. S., Di Donato N., Garavelli L., Grønborg S., Herkert J. C., Hoogeboom A. J., Jamsheer A., Latos-Bielenska A., Maat-Kievit A., Magnani C., Marcelis C., Mathijssen I. B., Nielsen M., Otten E., Ousager L. B., Pilch J., Plomp A., Poke G., Poluha A., Posmyk R., Rieubland C., Silengo M., Simon M., Steichen E., Stumpel C., Szakszon K., Polonkai E., van den Ende J., van der Steen A., van Essen T., van Haeringen A., van Hagen J. M., Verheij J. B., Mannens M. M., Hennekam R. C. Phenotype and genotype in 103 patients with trichorhinophalangeal syndrome. *Eur J Med Genet*. 2015 May, 279-292.

[4] Candamourty R., Venkatachalam S., Karthikeyan B., Babu M. R. Trichorhinophalangeal syndrome type 1: A case report with literature review. *J Nat Sci Biol Med*. 2012 July, 209-211.

[5] Maas S., Shaw A/, Bikker H/, Hennekam R. C. M. Trichorhino-phalangeal Syndrome. *Gene Reviews®*, 2017 Apr, 1993-2017.

[6] Shin H. T., Chang M. W. Trichorhinophalangeal syndrome, type II (Langer-Giedion syndrome). *Dermatol Online J*. (2001).

[7] Tsang W. K., Yang K. W., Fong C. M. Langer-Giedion syndrome: the evolving imaging features in hands and beyond. *Skeletal Radiol*. 43 (2014) 251-255.

[8] Riedl S., Giedion A., Schweitzer K., Müllner-Eidenböck A., Grill F., Frisch H., Lüdecke H. J. Pronounced short stature in a girl with trichorhinophalangeal syndrome II (TRPS II, Langer-Giedion

syndrome) and growth hormone deficiency. *Am J Med Genet A*. 2004 Dec, 200-203.

[9] Niikawa N., Kamei T., The Sugio-Kajii syndrome, proposed trichorhinophalangeal syndrome type III, *Am. J. Med. Genet*. 278-280.

[10] Solc Roman, Klugerova Michaela, Vcelak Josef, Baxova Alice, MiloslavKuklik, JanVseticka, Rastislav Beharka, Katerina Hirschfeldova, Mutation analysis of *TRPS1* gene including core promoter, 5′UTR, and 3′UTR regulatory sequences with insight into their organization., *Clinica Chimica Acta* 2017, 30-36

[11] Piccione M., Niceta M., Antona V. et al. Identification of two new mutations in *TRPS 1* gene leading to the trichorhinophalangeal syndrome type I and III. *Am J Med Genet A*. 149A (2009) 1837-1841.

[12] Lüdecke H. J., Wagner M. J., Nardmann J., La Pillo B., Parrish J. E., Willems P. J., Haan E. A., Frydman M., Hamers G. J., Wells D. E., Horsthemke B. Molecular dissection of a contiguous gene syndrome: Localization of the genes involved in the Langer-Giedion syndrome. *Hum Mol Genet* (1995) 31-36.

[13] Lüdecke H. J., Schmidt O., Nardmann J., von Holtum D., Meinecke P., Muenke M., Horsthemke B. Genes and chromosomal breakpoints in the Langer-Giedion syndrome region on human chromosome 8. *Hum Genet* 105 (1999) 619-628.

[14] Lüdecke H. J., Johnson C., Wagner M. J., Wells D. E., Turleau C., Tommerup N., Latos-Bielenska A., Sandig K. R., Meinecke P., Zabel B., Horsthemke B. Molecular definition of the shortest region of deletion overlap in the Langer-Giedion syndrome. *Am J Hum Genet* 49 (1991) 1197-1206.

[15] Chen C. P., Lin S. P., Liu Y. P., Chern S. R., Wu P. S., Su J. W., Chen Y. T., Lee C. C., Wang W. An interstitial deletion of 8q23.3-q24.22 associated with Langer-Giedion syndrome, Cornelia de Lange syndrome and epilepsy. *Gene*. 15 (2013) 176-180.

[16] Deardorff M. A., Wilde J. J., Albrecht M., Dickinson E., Tennstedt S., Braunholz D., Mönnich M., Yan Y., Xu W., Gil-Rodríguez M. C.,

Clark D., Hakonarson H., Halbach S., Michelis L. D., Rampuria A., Rossier E., Spranger S., Van Maldergem L., Lynch S. A., Gillessen-Kaesbach G., Lüdecke H. J., Ramsay R. G., McKay M. J., Krantz I. D., Xu H., Horsfield J. A., Kaiser F. J. *RAD21* mutations cause a human cohesinopathy. *Am J Hum Genet.* 90 (2012) 1014-1027.

[17] Kang S. H., Shaw C., Ou Z., Eng P. A., Cooper M. L., Pursley A. N., Sahoo T., Bacino C. A., Chinault A. C., Stankiewicz P., Patel A., Lupski J. R., Cheung S. W., Insertional translocation detected using FISH confirmation of arraycomparative genomic hybridization (aCGH) results, *American Journal of Medical Genetics*. Part A 152A (2010) 1111-1126.

[18] Nowakowska B. A., de Leeuw N., Ruivenkamp C. A., Sikkema-Raddatz B., Crolla J. A., Thoelen R., Koopmans M., den Hollander N., van Haeringen A., van der Kevie-Kersemaekers A. M., Pfundt R., Mieloo H., van Essen T., de Vries B. B., Green A., Reardon W., Fryns J. P., Vermeesch J. R., Parental insertional balanced translocations are an important cause of apparently *de novo* CNVs in patients with developmental anomalies, *European Journal of Human Genetics* 20 (2012) 166-170.

[19] Frasson M., Calixto N., Cronemberger S., de Aguiar R. A., Leão L. L., de Aguiar M. J., Oculodentodigital dysplasia: study of ophthalmological and clinical manifestations in three boys with probably autosomal recessive inheritance. *Ophthalmic Genet*. 2004 Sep, 227-236.

[20] Matesic D., Hagan J. B. Cartilage-hair hypoplasia. *Mayo Clin Proc*. 2007 Jun.

[21] Baujat G., Le Merrer M. Ellis-van Creveld syndrome. *Review. Orphanet J Rare Dis*. 2007 Jun.

# INDEX

## A



## B



## C

## D



## E

## K



## L



## M



## N



## O



## P

## R



## S



## T



## X



## Y